Camille Fauré

Limoges Art Deco Enamels

The Geometry of Joy

Camille Fauré

Limoges Art Deco Enamels

The Geometry of Joy

Alberto Shayo

ANTIQUE COLLECTORS' CLUB

To the three graces in my life, my sisters Arlette, Tounée and Rita

---

First published 2007

ISBN 978-1-85149-535-1

Frontispiece, vase, see page 215
Endpapers, detail from vase, see page 118

British Library Cataloguing-in-Publication Data:
A catalogue record for this book is available from the British Library

Printed in China for The Antique Collectors' Club Ltd.,
Woodbridge, Suffolk IP12 4SD

Contents

Acknowledgements

Writing a book is no simple task but it becomes easier when you have good people on board:

With Clara Sancovsky on your side any venture is a guaranteed success. Her knowledge is unrivalled and her enthusiasm is only surpassed by her charm. I am deeply indebted to her.

Sharon Kerman has travelled with me doing research work. Her acute investigative knowledge and persistent manner has guided me through the fantastic world of enamels.

Diana Steel, my publisher, for being highly professional, warm and ever so patient with me.

Monsieur and Madame Robert Lavaurs have been magnificent hosts, opening family records that had never seen the light. The artist came to life through their efforts.

Pierre and Laurent Lavaurs aided in the compilation of facts and images.

Monsieur and Madame Georges Lavaurs volunteered specific details helping to round off loose ends.

Madame Françoise Malabre appeared one day with bagfuls of archival data and magnificent pieces. I could not believe my eyes.

Monsieur and Madame Jean-Michel Malabre welcomed me and spoke at length of family matters. Their collection is of particular interest since it includes prototypes and experimental pieces. They are all illustrated.

Madame Véronique Notin, head of the Musée Municipal de l'Evêché in Limoges, was most forthcoming in the information I requested; she is well versed on the subject. Thanks are also due to her assistant Monsieur Alain Dionnet.

Bernard Vergne, a most charming gentleman, gladly opened up the records of Edmond Allain and more. Still in the same business as his grandfather, he has all the records one could ever hope for.

Paul Buforn, 'maître émailleur' in Limoges, a former artisan at Fauré, with his soft manner was instrumental at the beginning and end of this project with specific information that opened many doors.

Alain Grafeuil, 'maître émailleur' in Limoges, his archives were eye openers. His help was priceless.

Mauricette Pinoteau, 'émailleuse' in Limoges, it was simply a pleasure to sit and hear her recollect her years at Fauré. Charm and knowledge go hand in hand in this artist. Probably the only artist today to retain some of the spirit of the Fauré style.

Michel Bétourné, 'émailleur' in Limoges, who so candidly recollected memories of his father Raphaël Bétourné.

Christian and Simone Christel, 'émailleurs' in Limoges, who recollected the day-to-day life of the enamel milieu. Today they stand as the avant-gardists of the enamel tradition.

Andy Thomain, with his knowledge of Limoges and its collections; his goodwill and collaboration are greatly appreciated.

Professoressa Annamaria Andreoli and Mariangela Calubini, at the Vittoriale in Gardone Riviera in Italy, never hesitated to give me the most useful revised data.

Franck Dufour and Micheline Lasserre, at the Cristallerie Saint Paul, gave me the grand tour of the factory and showed me how enamel, the raw material, is made.

Maître Philippe Urbain, for the legal documents relating to the liquidation of the Fauré enterprise.

Monsieur Jacques Ladde, 'directeur opérationnel de la SEPEL', and Madame Monique Barriol in Lyon, who revived for me the 'Foire de Lyon'.

Alastair Duncan, always supportive in my endeavors.

Marcio Roiter, an ever present ally in all my projects.

Evelyne Possémé, curator of the Art Deco-Art Nouveau department of the Paris Musée des Arts Décoratifs, was very helpful in providing information.

Estelle Fallet, at the Musée de l'Horlogerie et de l'Emaillerie de Genève.

Odette Lajudie, the dedicated secretary of the atelier Fauré. Her insight prevails on many subjects.

Danielle Soury McCann, guide at the atelier Fauré between 1965 and 1968, and devoted assistant to M. Gérard Malabre when he was President of the Biennale.

Simi Moryoussef. Charm and experience make up this wonderful person.

And last, but never least, Marcelo Ferrari, a bright, upcoming photographer and technical expert who guided me through the jungles of computers.

The many who wish to remain anonymous.

Preface

This is my third book. The excitement and expectation is no less than for the others.

The anguish of an author can be the fuel to an invigorating novel; and the happiness of another author can be the desire to investigate without fear of falling. I fall in the second category. If happiness is a state of mind, my entourage keeps me where I am. Yet again, I am in a good place.

Just as a ripe apple can do nothing but fall off a tree, my spirit was set to plunge into the world of enamels, of Art Deco enamels. The work of Camille Fauré has been familiar to me for decades. It was time to encounter the man and unravel the aura that surrounds him.

It is very hard not to be seduced by the allure and magic of the enamels of the Art Deco period, particularly by the art that came out of the Fauré workshop. Its hard, shiny surfaces with vibrant colours and deep sculpted motifs, whether blooming flowers in their brazen natural state or detailed geometric patterns that intertwine themselves in repetitions, are always a pleasing experience.

In France, Limoges has been the centre of enamelling for centuries. I visited Limoges on several occasions, encountered its porcelain, its enamellers, its raw materials and, above all, I met its artists.

For Fauré, workmanship was the most important factor in the atelier. Great effort was given to detail, composition and beauty. Time was of no importance. The end product was paramount.

Today the Fauré aesthetic may no longer be there. With the passage of time, styles change, demand changes, workmanship changes and the outlook on the future of its art changes too.

Yet new ideas spring out; many are based on centuries-old roots and the influence that the Art Deco period left on the city is unequivocal. A walk on its streets and a look into its architecture and its galleries whether exhibiting enamels or porcelain is all the evidence needed.

Enjoy this 'joy'.

Alberto Shayo

Camille Fauré as a young man

CAMILLE FAURÉ

Limoges Art Deco Enamels

The Geometry of Joy

Camille Fauré was the founder and director of a workshop that produced some of the finest enamelled art objects of the Art Deco period. The superb quality of the Fauré creations gave the firm an aura that has lasted for decades. Yet Camille Fauré never designed or created a single piece himself.

Camille Fauré enamels are the best known of the Art Deco period, and even beyond, because they embody the era's vibrancy and audacity. The hard, shiny surface of enamelled metal, the intensity of its colours, and the technique of sculpting volumes and geometrical forms in the enamel itself – a technique for which the Fauré workshop was famous – were all welcomed enthusiastically as an expression of the vitality of Art Deco.

Production was prolific, and today Fauré pieces can be found all over the world. They are remarkable for their stunning variety of design and decorative technique, as well as for their level of quality, which remained unsurpassed, particularly during the period from the mid-1920s to the end of the 1950s. If a piece was allowed to leave the Fauré workshop and be offered for sale, one could be sure that it was absolutely perfect; even a minor defect would cause an object to be rejected by the Atelier.

Early years

Camille Raymond Fauré was born in Périgueux on 16 November 1874. He was the son of Jean Marie Bertrand Fauré (born c.1840, Toulouse - died 1895, Limoges), a 34-year-old painter – as it is stated on the baby's birth certificate – and his 32-year-old wife Marie Pinsan (b.1845 - d.1924). The couple had married in Bordeaux and were living at 29, rue Cyrano in Périgueux when Camille was born.

The couple had two other children, daughters named Blanche and Madeleine, both younger than Camille. In later years, Blanche, who never married, had a chocolate shop in Paris' rue du Faubourg St. Honoré called 'La Pitchounette', where she sold chocolates that were sent specially from Limoges. Madeleine became a teacher and a school principal, and married another teacher and school principal named Léonard Lajoie.

Though little is known of the family's early life, it is certain that by the early 1890s the family was living in Limoges. In 1892, a city directory (the *Almanach Annuaire*) lists: "Fauré, peintre en bâtiment" [house painter] and "Fauré, lingère, repasseuse" [seamstress], at place Etienne Pinchaud, 7. Thus we learn that Jean Marie Fauré was working as a house painter, no doubt assisted by his son Camille, then 18, and that his wife Marie (sometimes also called Melanie, probably a second name) was contributing to the family income by ironing and mending. This listing continues unchanged until the death of Jean Marie on 7 April 1895, in his mid-fifties. The two witnesses listed on his death certificate were François Fauré, his brother, also a painter, and Léonard Lajoie, his son-in-law.

At the time all young men had to register for military service when they turned twenty. Camille Fauré's name is thus duly recorded in the military archives in Limoges under the matriculation No. 637. The records show that, although he registered, as was compulsory, in 1894, and passed before an examination board both that year and the following year, he did not actually serve until 1897. A word scrawled in the register in both 1894 and 1895 may explain the exemption during those two years: "faiblesse" [weakness]. This is admittedly a rather vague criterion, and other possible reasons for exempting him are listed: he was a skilled artisan ("ouvrier d'art") and, by 1895, the only son of a widow. In November 1897, Camille Fauré was incorporated into the 63rd regiment of the Infantry as a second-class soldier, and carried out his military service for the prescribed amount of time. As this was before the generalised use of photography for identification purposes, the physical appearance of each soldier was sketched out verbally, in compliance with set criteria. According to Camille's description, he had brown hair and eyebrows, an "ordinary" forehead, a "medium-sized" nose and mouth, a round chin, and an oval shaped face.

The records confirm Camille's address at the time: it is noted that he lived in Limoges at 7, Place Etienne Pinchaud and exercised the profession of "peintre décorateur en bâtiment" [house painter and decorator].

While the laws promulgated by Jules Ferry had made public schooling universal, free, and obligatory in France in the 1880s, school for all was still a relatively new concept. Part of the military examiners' job was to evaluate the level of instruction of the young men who presented themselves before them. Camille Fauré's record is quite respectable: he knew how to read and write, he had his certificate of primary studies ("certificat d'études primaires"), and he played the clarinet – this being perhaps pertinent at a time when military fanfares could still be counted on to boost the morale of the troops. Additional notations mention that he did not know how to swim, to ride, or to care for a horse, but was a certified cyclist ("breveté vélocipédiste").

Fauré's first stint in the military ended on 18 September 1898, when he was discharged ("envoyé dans la disponibilité") and retired to Limoges.

Striking out alone

After he finished his military duty, Camille continued to exercise the trade he had learned alongside his father. In 1900 the *Almanach Annuaire* states that he was a painter and living with his mother at 29, rue des Combes.

The first letterhead stationery we have for him is at the address of 29, rue des Combes and 4, Place Fontaine des Barres (the building being situated at the junction of the two streets). The place Fontaine des Barres is, even today, a picturesque little square with a 17th-century pyramid-shaped fountain and lovely, half-timbered houses. A letter written on this early stationery and dated 1904 has been conserved by the family. It shows that the young enterprise was quite eclectic.

The next change in address comes in 1903, when Fauré is listed in the Dumont city directory at 30, Avenue Garibaldi. This is a move that would have made Fauré's establishment much more visible; the avenue Garibaldi is a wide and busy avenue that was home to several other enamellers and artisans, among them Jules Sarlandie (No. 13) and, later on, Alexandre and Henriette Marty, (No. 52 bis). The entry in the directory announces an important change in the firm's orientation. It says: "Fauré, peintre d'enseignes" [sign painter] Avenue Garibaldi 30.

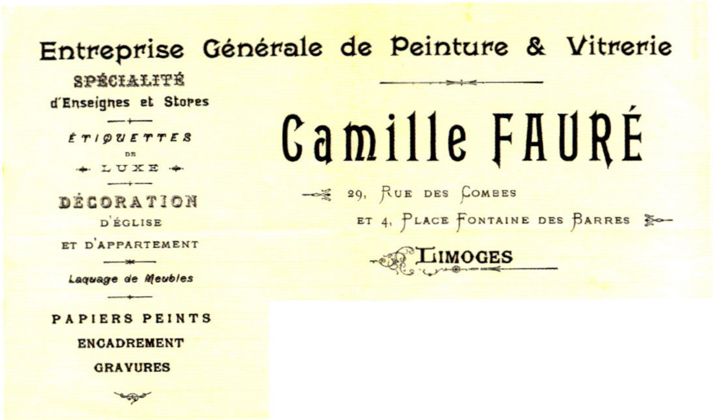

Letterhead of Camille Fauré works at rue des Combes, 1904. It lists the skills and services on offer from his "General Painting and Glazing Enterprise": "Specialising in Signs and Blinds; Luxury Labels; Decoration of Churches and Apartments; Lacquering of Furniture; Wallpaper; Framing; Engraving"

Letterhead of Camille Fauré works at Av. Garibaldi, 1905

In the company's new letterhead, of which we possess a specimen dated June 1905, the services offered are much the same, except that the list of specialities is shorter and a new proficiency has been added: that of trompe-l'oeil decoration in faux wood and marble.

These early years, during which he was trying to establish his business, were not easy ones for Camille. During this time he often sought the advice of his brother-in-law, Léonard Lajoie. Lajoie (he preferred to be called by his surname) lived with his wife Madeleine, Camille's younger sister, in the neighbouring town of Château-Ponsac. The two men were very close, almost like brothers.

Examples of lettering for signs. Silver and gold leaves are used to achieve desired effects

Pharmacy sign designed by Fauré as it stands today at the corner of Av. Garibaldi and Av. de la Libération in Limoges

Very quickly, the main business of the establishment at 30, Avenue Garibaldi became that of sign painting. All sorts of signs and plaques were furnished, initially for advertising various firms, and for railway and tramway insignias. Later, the firm also furnished street signs and gravestone plaques. These were executed in various ways: on copper, on glass, with gold leaf, and featuring many different styles of lettering.

A sign painted by the Fauré Company, though at a later date, may still be seen in Limoges at the corner of the Avenue Garibaldi and the Avenue de la Libération. Proclaiming the word "Pharmacie" in gold letters on a black background, the massive, wood-framed sign is now registered as part of the city's artistic heritage and, as such, may not be destroyed.

For many years Fauré's foreman in the sign painting enterprise was a man named Pierre Bouleisteix. Fauré relied on Bouleisteix to oversee the daily running of the business while he sought out customers. The two men were on very friendly terms, Bouleisteix even calling his boss by his first name and addressing him with the familiar "tu" form.

The early 20th century, which saw the rise of department stores, veritable palaces dedicated to consumer goods, and expanding transportation networks, also saw a growth in the demand for advertising. Soon the Fauré sign-painting business branched out to include signs and plaques executed in enamel. Much in demand from the early years of the century onward, these plaques were extremely sturdy due to the hardness of enamel, and could be attractively decorated with colourful slogans and eye-catching motifs. Thus, for a while, the Fauré business consisted of two branches: house painting, decoration, and sign painting; and enamel advertising plaques.

It was in 1912 that the first mention of Fauré's most important address appeared in the Dumont directory: "Fauré, Camille, peintre d'enseignes [sign painter], 31, rue des Tanneries". From that moment on, a portion of the Fauré business would always be located in the rue des Tanneries.

By 1920, the company was no longer listed at No. 30, but at No. 61, of the Avenue Garibaldi. This workshop was now largely, if not exclusively, devoted to the production of enamel signs. The two listings in the Dumont directory for this year were: "Fauré, Avenue Garibaldi, 61, Emaux de Limoges" [Limoges Enamels] and "Fauré, rue des Tanneries, 31, Peinture d'enseignes" (sign painting]. Under the latter heading, an attractive advertisement in the Art Nouveau style is inserted:

Sign Manufactory
Camille Fauré
31, rue des Tanneries
All letters and signs, labels, enamel plaques, decoration.
We are the only manufacturer of these goods;
our items are cheaper and better made.

A letterhead from this period for the Avenue Garibaldi confirms the growing importance of enamel for the firm:

Emaillerie de Limoges
Panneaux émail pour enseignes, attributs et publicités
[Enamel plaques to serve as signs, for identification, and as advertising]
Camille Fauré et Cie.
Usine et bureaux: 61, Av Garibaldi, Limoges
Téléphone 10 30.

Camille Fauré (standing in the foreground) during World War I, in which he served as a railway guard

Family life

At the same time that Fauré was establishing his businesses, he also met the woman who was to become his wife. Marie Madeleine Barbaud, called Madeleine like Camille's younger sister, was born in Limoges on 16 August 1883, to Bertrand Barbaud, a forty-one-year-old stone cutter, and his wife Jeanne Marie Roch, then twenty-three.

Camille and Madeleine were married in Limoges on 31 May 1900. According to the marriage certificate, the bridegroom, an "entrepreneur", was twenty-five and the bride just sixteen. The certificate further states that at the time Camille was living with his mother at 29, rue des Combes, and Madeleine with her mother – her father having died in 1898 – in the rue de Paris, 32. The bride being under age, her mother had given her consent to the marriage. Among the witnesses was Camille's brother-in-law and close friend, Léonard Lajoie.

The couple had two daughters, Réjane, who was born in 1902, and Andrée, born on 3 December 1904. The family's artistic leanings were already apparent: their first daughter was named for Gabrielle Réjane, a much loved actress in turn-of-the-century France. Réjane became a gifted pianist, studied at the Limoges conservatory and had planned to become a piano teacher before her marriage. Both Réjane and her sister Andrée later attended the Limoges School of Decorative Arts, where Andrée, in particular, was an excellent student.

Music was important in the Fauré family. Camille Fauré is remembered as being an excellent musician and a lover of opera, who liked to follow the score while listening to music. Though none of the family recalls ever seeing him with a clarinet, (contrary to the statement made by the military examiners in 1894), they are unanimous in praising his singing voice. Camille Fauré would often calm his baby granddaughter's crying by singing to her (that baby, Florence, later became a pianist and teacher at the Marseille conservatory). It has been said that the composer Gabriel Fauré (1845-1924) and Camille are related, and judging from the latter's musical gifts, this may well be true. As yet, however, no evidence has been found to prove the assertion.

When World War I broke out, Fauré was called back to military service. Probably due to his age – by this time he was forty – he was not sent for active duty. The family recalls him guarding the railway system, not an unpleasant duty. Indeed, a photo from this period shows a boyish and jovial Camille standing with a group of fellow soldiers and looking as carefree as if he were on a picnic.

In 1917 he was assigned to the division responsible for automobiles and vehicles ("sert au parc automobile"). In early 1919 this second tour of duty came to an end, and Camille returned home, to 31, rue des Tanneries on 11 January, "alone and having received the sum of zero francs for his expenses", as the military records duly note.

Rue des Tanneries

The rue des Tanneries, where Camille Fauré was first listed in 1912, was to be the centre of his professional activities for the rest of his life, and would remain the Fauré company address until the end.

In 1914, an additional – and short-lived – address in the Boulevard de Fleurus, as well as novel activities, are listed in an advertisement in the Dumont directory:

> *Painting, Glazing, and Wallpapering*
> *Artistic Decoration in all Styles*
> *Camille Fauré*
> *Rue des Tanneries - Bd. de Fleurus*
> *Telephone: 9 - 47*
> *The company is known for its fine workmanship*

Detail of the mosaic tiles in the rue des Tanneries outside the Fauré workshop

In 1924, the directory lists "Fauré (Camille) sign painter, rue des Tanneries, 31"; alongside "Fauré et Marty, art enamels", at the same address.

In 1935 construction began on a garage and storeroom in the rue des Tanneries, in order to house the expanding enamel workshop on the same footing as the sign-painting workshop. The enamel works were installed on the left-hand side of the premises, and the sign painting on the right-hand side. During the remodelling, Camille spent much of his time in the country. It was at this time that the furnace for the enamels was installed in the basement.

Around 1937 the building in the rue des Tanneries, which still stands today, was given an Art Deco style façade featuring a double band of gold and brown iridescent mosaic tiles set in a checkerboard pattern that is reminiscent of the iridescent colours of Fauré enamels.

Some time later, further remodelling resulted in offices being built in the courtyard, as well as an apartment that housed Andrée Fauré Malabre and her family. Thus, for many years, the Malabres lived in the same building as Camille Fauré, whose apartment was on the first floor, and just above the enamel workshop.

In the late 1950s, the street numbering was changed, and No. 31 of the rue des Tanneries became No. 41.

LIMOGES AND ENAMEL

Enamel workshops were in existence in Limoges by the end of the 12th century. The primary technique was *champlevé* enamel, in which a portion of the metal support is removed forming a slight hollow into which the enamel is applied.

At the end of the 15th century in Limoges, the technique of painting on enamel again brought Limoges to the forefront of the art, with such masterful practitioners as the Limousin, Reymond, and Pénicaud dynasties.

During the 16th century, Limoges was the leading enamelling centre in Europe; one of the foremost techniques being that of *grisaille*, in which an image is rendered in shades of one colour, usually grey or white, over a darker background.

Enamelling was rediscovered during the 19th century, thanks to a growing interest in past art forms. During the early 20th century, artists such as Paul Bonnaud (1873-1953); Jules Sarlandie (1874-1936); and Léon Jouhaud (1874-1950) brought innovation and mastery to enamelling. The Art Deco style was an immediate success, with the rise of the workshops of Camille Fauré (1874-1956) and his daughter Andrée (1904-1985), and Alexandre Marty (1976-1943) and his daughter Henriette Marty (1902-1996).

The more independent post-war generation of enamellers considered enamel as a means of personal expression. New names appeared in the world of Limoges enamelling: Georges Magadoux, Christian Christel, Roger Duban, Boris and Catherine Veisbrot, among others. After the Fauré workshop closed, many of its employees opened up shop on their own, and are still active now.

Today's enamellers often find it difficult to survive. The public, unaware of the time-consuming nature of enamelling, often consider the prices high. Though the enamel powders are of better quality, many colours are no longer available, due either to prohibitive cost or to the toxicity of their components. Still, the devotion of these artisans keeps the art alive.

Camille Fauré, the man

Camille Fauré was a life-long radical-socialist. He was a good friend of Limoges' socialist mayor, Léon Bétoulle. Bétoulle, the illegitimate son of a seamstress, had worked his way up through the ranks to become mayor of the city. It was a position he held practically all his adult life, from 1912 until his death in 1956 (he and his friend Camille Fauré died in the same year) with only a brief interruption from 1941 to 1947, due to the war and its aftermath.

Though, politically, his heart may have been on the left, Camille Fauré enjoyed living like a country gentleman and a "patron", a proprietor. He was a born manager, with a global vision and an unerring sense that helped him choose the best and most skilled work associates. This combination is perhaps less unusual in Limoges, where even the middle classes traditionally tend to be left-wing and anti-clerical. "In Limoges," said a member of the Fauré family, "the bourgeois are all radical-socialists".

By all accounts, Camille Fauré was a *bon vivant* who took advantage of life's pleasures. He loved cycling in the countryside, and enjoyed many sports, with a special fondness for fishing and hunting. Though he never played, he was a rugby fan and was President of the Limoges rugby club from 1939 to 1945. His sons-in-law Louis Malabre and Pierre Lavaurs did play, as did his grandsons and great-grandsons. Camille first met Louis Lavaurs, his son-in-law's father, at a rugby event. He never missed a game, and though he loved going out to the country, would never go when one was scheduled.

A tall and handsome man, Camille was always impeccably dressed in a suit and tie. He never wore the workman's smock, even when in the workshop, and had an innate elegance that singled him out. The twinkle in his eye and the upturned moustache that can be seen in all his photographs mark him as a man who enjoyed life and intended to do as he pleased. He made a point, in any case, of cherishing his 'bad' habits: though his wife always asked him to cut down, he continued to smoke all his life and he was an inveterate card player.

Every afternoon Camille would go to the café across the street from his apartment – the Café Riche, in the Place de la République – to play *manille* with his cronies; and every evening around 7pm, his wife would send his young grandson to fetch him. That grandson, Robert Lavaurs, today a grandfather himself, remembers the exact words he used to say, with the affectionate names he called his grandparents: "Pépé, Mané says you should come home. The soup is ready."

Camille Fauré, the businessman

While in the early days Camille Fauré was certainly extremely active, in later years, once his businesses were on a firm footing, he was quite content to give a free rein to those he had put in charge of them, and in whom he trusted completely. Even then, though he was not a practitioner and never wielded a paintbrush or a spatula in his life, in his own way Fauré continued to take an active part in the daily life of his two workshops. Every day, in the rue des Tanneries, he would rise in the mid-morning, breakfast and dress, and come down around 11 o'clock, making the rounds first of the sign-painting workshop and then of the enamel workshop. Always dapper and affable, he was liked and respected by his employees, with whom he would chat and keep himself informed on the latest happenings in the workshop. As one of them recalled: "He never spoke an unpleasant word". A talented manager with a flair for sensing talent in others, Camille Fauré gave his workers sufficient liberty to develop their abilities, while tactfully, yet energetically, maintaining the high standards of perfection that characterised the atelier.

Signs and enamel

For several years, the Fauré sign painting and enamel advertising businesses flourished side by side. Initially, there were more sign painters than there were enamellers, for in those days advertising depended on huge, hand-painted signs that were frequently changed. This required many hands and involved a great deal of labour. During the 1940s, one of the company's major activities was to furnish the signs advertising the latest films, with the names and faces of the stars, which were still hand-painted at that time, before posters came into general use.

The recollections of former employees of the sign-painting business depict a relaxed, family-like atmosphere, whose structure was adaptable enough that people would occasionally even shift from one aspect of the business to another.

This was true for Michel Coiffe, who started working for Camille Fauré at the age of 15, in 1935. He began by helping out with the construction then taking place in the rue des Tanneries, where remodelling was being done. Coiffe was called to serve in the Second World War. When he returned, Camille Fauré had no need for another employee, and so sent him over to sign painting, then under the direction of his son-in-law Pierre Lavaurs.

Another example is that of Lucie Dadat, later one of Fauré's finest Art Deco enamellers, who began by wielding a spray-gun in the sign-painting shop – a far cry from the delicate precision of the enamel spatula!

Outside the Fauré workshop, 1925.
Workers in front of Fauré billboards on Rue des Tanneries at Limoges

Partial view of a cinema billboard created and signed by the Fauré workshop. The enamel workshop is just behind

In both workshops, the day began at 8:30 in the morning, ending at 6:00 in the evening for the sign painters and 6:30 for the enamellers. There was a two-hour break for lunch from 12 to 2pm, during which everyone went home to eat, Limoges being a small city and the employees all living nearby.

Raphaël Bétourné

An entrepreneur at heart, Camille Fauré soon realised that just as sign painting had given him the opportunity to branch out to the production of enamel advertising plaques, the gap that separated enamel advertising plaques from the decorative arts was a small one that could easily be bridged.

In order to do so, Fauré, always quick to appreciate the expertise and abilities of others, sought help. He got in touch with an enameller then working in Paris, Raphaël Bétourné. Bétourné, who had learned his trade in Paris with Lucien Delrive – a student of Alfred Meyer – had opened a workshop there, and was working there when World War I broke out. The two were introduced by Louis Clément, a Limoges enameller who would later become Bétourné's associate. Fauré offered Bétourné a position in his new art enamel workshop.

After an initial hesitation, Bétourné accepted the offer. In doing so, his son Michel recalls, he was probably influenced by the German bombardment of Paris in 1918. The event was a terrible shock for him, and made him resolve to leave the city. Thus, Bétourné went to Limoges and began his working relationship with Fauré around 1919. Though it would last only a year, it was to mark a decisive turn for Fauré.

From that time onward, the Fauré art enamel workshop would continually grow in importance, taking on a momentum all its own. Upon leaving Fauré in 1920, Raphaël Bétourné opened his own workshop, initially in Limoges' Avenue de Bénédictins, and for a time in association with Clément. In later years Bétourné's sons Michel and Jean each had a workshop; Michel's being in the rue des Tanneries, very close to that of Fauré.

By 1923, Fauré had left his workshop in the Avenue Garibaldi altogether and had consolidated both of his enterprises in the rue des Tanneries.

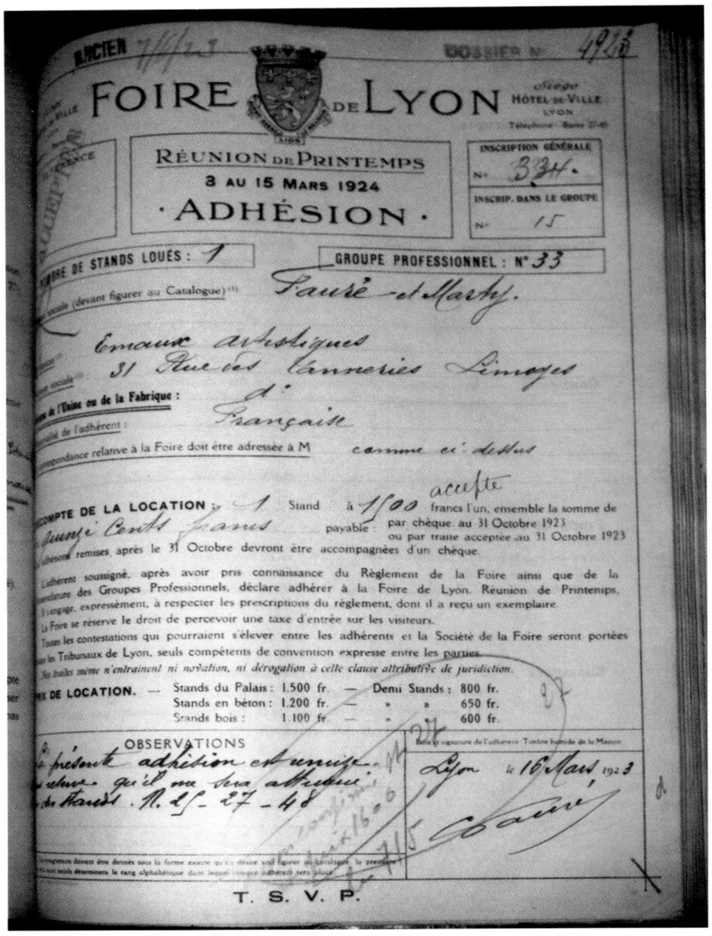
FOIRE DE LYON
Siège Hôtel de Ville Lyon
RÉUNION DE PRINTEMPS
3 AU 15 MARS 1924
· ADHÉSION ·
INSCRIPTION GÉNÉRALE N° 334.
INSCRIP. DANS LE GROUPE N° 15
STANDS LOUÉS : 1
GROUPE PROFESSIONNEL : N° 33
(devant figurer au Catalogue) Fauré et Marty.
Emaux artistiques
31 Rue des Tanneries Limoges
de l'Usine ou de la Fabrique : d°
de l'adhérent : Française
…dance relative à la Foire doit être adressée à M comme ci-dessus
…PTE DE LA LOCATION : 1 Stand à 1500 francs l'un, ensemble la somme de Quinze Cents francs payable : par chèque au 31 Octobre 1923 ou par traite acceptée au 31 Octobre 1923
… remises après le 31 Octobre devront être accompagnées d'un chèque.
…adhérent soussigné, après avoir pris connaissance du Règlement de la Foire ainsi que de la …ture des Groupes Professionnels, déclare adhérer à la Foire de Lyon, Réunion de Printemps, …engage, expressément, à respecter les prescriptions du règlement, dont il a reçu un exemplaire.
La Foire se réserve le droit de percevoir une taxe d'entrée sur les visiteurs.
…les contestations qui pourraient s'élever entre les adhérents et la Société de la Foire seront portées …les Tribunaux de Lyon, seuls compétents de convention expresse entre les parties.
…traites même n'entraînent ni novation, ni dérogation à cette clause attributive de juridiction.
…X DE LOCATION. — Stands du Palais : 1.500 fr. — Demi Stands : 800 fr.
Stands en béton : 1.200 fr. — » » 650 fr.
Stands bois : 1.100 fr. — » » 600 fr.
OBSERVATIONS
…présente adhésion est … qu'il ne sera attribué … N. 25 - 27 - 48
Lyon le 16 Mars 1923
Fauré
T. S. V. P.

Fauré Marty application form for the Lyon fair 1924

Camille Fauré's daughters, Andrée (left) and Réjane, in 1926

Fauré et Marty

Another colleague took Bétourné's place in the newly-created enamel workshop. This was Alexandre Marty, who had been trained as a porcelain craftsman and practised both crafts, like many others in the field of enamelling. Marty (1876-1943) would also soon strike out on his own, working together with his daughter Henriette (1902-1996) to create exceptional pieces in the Art Deco style.

It is generally considered that Marty played the rôle of catalyst, introducing Fauré to the possibilities offered by the new Art Deco style. Many of the distinguishing characteristics of Fauré's Art Deco pieces probably originate from the Fauré Marty partnership. The workshop also received a great impetus from the arrival of one of Fauré's most gifted enamellers, Pierre Bardy, in 1919. (He was to remain there for 42 years.) The effervescence resulting from the combination of all these talents was crucial to the development of a new Art Deco enamel vocabulary.

Although the association between Fauré and Marty was short-lived, lasting only from the early 1920s to 1925, it was intense and fruitful. A number of pieces are known to exist with their joint signatures. Fauré and Marty exhibited together at the Foire de Lyon from 1922 to 1925. Their application for the 1922 fair, apparently their first showing as a partnership, is dated 22 September 1921. It gives their business name as "Fauré et Marty", and states that they deal in "art enamels". In 1924, the listing of the Fauré Marty company in the Lyon Fair directory provides interesting information on their production at the time: "Fauré et Marty, Emaux d'Art, tous modèles pour bijoux, vases artistiques, modèles exclusives, tableaux anciens et modernes, Limoges". [Fauré and Marty, Art Enamels, all sorts of models for jewellery, artistic vases, exclusive models, old master and modern paintings, Limoges.]

The 1925 Lyon Fair was the last that Fauré and Marty attended together. In 1926, Fauré again exhibited alone, as "C. Fauré" (the alphabetical list of that fair, however, continued to mention "Fauré et Marty", but this was probably just a hold-over from previous years). The association was most likely dissolved shortly after that. In 1926, Marty appeared in the *Tout Limoges* directory alone, at 6, Impasse de Fontaury.

Fauré's daughters and their families

In 1920, Réjane married Pierre Lavaurs, whose family was also in the business of sign painting in Villefranche, and whose father, Louis Lavaurs, Camille had first met at a rugby game. Réjane and Pierre had two sons, Robert and Georges; the former eventually took charge of the sign business, which in due course turned into a contemporary advertising business that has

The Lyon Fair in 1934

remained in the family to this day, while Georges became a renowned cardiologist in Marseille. After the Second World War, Réjane Fauré Lavaurs became a Limoges City Councillor under Mayor Georges Guingoin, who was mayor from 1945 to 1947. Réjane died in 1968, of a heart attack. She and her sister had been very close all their lives, speaking every day on the telephone even though they lived quite near to each other.

Andrée married another rugby player, Louis Malabre; their son Gérard was born on 28 July 1929. Though originally a shoe salesman, Louis soon placed his talents at the disposition of the Fauré establishment. A handsome man and an elegant dresser, he was also a dynamic and convincing salesman with a nearly irresistible charm. He could sell whatever he liked, and often made a game of it, setting himself challenges to test the strength of his powers of persuasion. Some time after his marriage to Andrée, Louis Malabre began working in the sign-painting portion of the business, eventually shifting to the enamel branch around 1937.

Louis Malabre very quickly became a key member of the Fauré team, searching out new clients and markets, and creating the commercial structure that Fauré had previously lacked. With time, Camille Fauré increasingly left the business in the capable hands of his daughter and son-in-law to devote himself to his favourite pursuits, hunting and fishing.

The Lyon Fair

The success of the Fauré art enamels, production of which had begun around 1919, led Fauré to seek out new commercial outlets. In 1920 he first began exhibiting at the Foire de Lyon, which was to be his main showcase for many years. Orders received at the Lyon Fair – and in later years, at the Paris Fair – were sufficient to maintain the company and furnish its twenty or so employees with enough work for a year at least. Among the purchasers were buyers for large gift shops, and the more elaborate and expensive an item, the better it seemed to sell.

Invitation for the 1930 Lyon World Trade Fair

The interior of the Lyon Fair in the mid-1940s

The Georges Cheyron stand at the Lyon Fair in 1951. Located in Paris, Cheyron provided the lampshades for the Fauré lamps for decades

The Fauré stand at the Foire de Paris in 1955

Fauré exhibit at the Lyon Fair in 1959

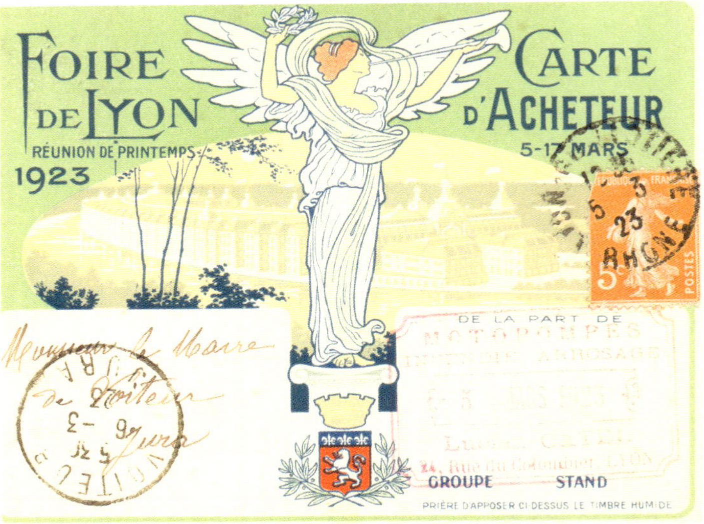

Buyer's card for the 1923 Lyon Fair. Buyers were required to carry and show this card whenever making purchases

Commemorative stamps for the Lyon Fair. Top: 1920; bottom left: 1933; bottom right: 1935

Advertisement for the Lyon Fair, 1931

This annual springtime event, the largest commercial outlet in France's Rhône-Alpes region, gave Fauré exposure to a wide clientele. The fair, held every spring, drew a large number of commercial representatives from all over the world, as well as a cross-section of the general public. Its brochure was printed in fourteen languages.

By 1920, the fair, which had begun in 1916 with 1,342 participants and a global turnover of 95 million francs, had grown to accommodate 6,000 participants, with a turnover of 1,500 million francs. The Fauré company would prepare for the event months beforehand, making a large number of pieces – between 250 and 300 approximately – especially for the fair, where they always sold well and the company made important business contacts.

Fauré first signed up for the Lyon Fair in November 1919,

Advertisement for the Lyon Fair, 1932

C. Fauré stand at the Lyon Fair, 1930

five months in advance. His application lists his business name as "Camille Fauré", giving the address of both the office and the workshop as 31, rue des Tanneries. Interestingly, that year he was listed under "Group 50: Jewellery". By 1922, the enamellers would be assigned to Group 33, which included ceramics, crystal, earthenware, porcelain, pottery, and glass – but still with no specific mention of enamel. In his 1919 application, Fauré described his activity as "Emaux d'Art de Limoges" [Limoges Art Enamels] and requested a blueprint of the stand so that he could prepare its decoration.

The account books of the Allain establishment of Paris, where the Fauré workshop bought its copper vase forms, mention vases bought by Fauré in January 1931, for the Lyon Fair.

Around 1955, Fauré also began showing at the Foire de Paris, and some years later he stopped attending the Lyon Fair. He had, however, attended the event faithfully, and profitably, for about forty years, at one point receiving an award for being among the oldest exhibitors at the fair.

January 1931		(F.F. = French Francs)
20.1.31	5 vases Yvonne at	36.30 F.F.
	5 vases Azyadée renversés	52 F.F.
In enamelling copper (containing 1% zinc)		
23.1.31	4 vases Artonio, h.87cm., Foire Lyon '31	45 F.F
	1 vase Yvonne	36.30 F.F.
27.1.31	4 vases Emile, h.195mm., Foire Lyon '31	38 F.F.
28.1.31	2 vases Amélie, h.30cm., Foire Lyon '31	87 F.F.
29.1.31	2 vases Lucien, h.290mm., Foire Lyon '31	54 F.F

This made a total of 23 vcses and 1,091.80 francs for that month.

Camille Fauré c.1930

The Art Deco Exhibition

Originally planned for 1914, the exhibition was postponed due to the war. The intervening years had been harrowing, and had changed the world forever. The brutality of the war, the shock of the trenches and poison gas, and the liberation of the women – who, in the absence of their husbands, fathers, and brothers, had been forced to take on unprecedented responsibilities and had found that they liked it – were all factors that led to a startlingly new society.

During the roaring twenties – in French, "les années folles", or the wild years – people sought to live with a vengeance, to forget the horrors and privations of the war. They delighted in the streamlined elegance, energetic geometrical lines, and intoxicating newness offered by the Art Deco style. With the 1925 exhibition, France sought to prove that it was still the supreme master of style, able to dictate what was fashionable. In the process, a style was named; "Art Deco" comes from the exhibition's title: "l'Exposition Internationale des **Arts Déco**ratifs et Industriels Modernes". (Interestingly, the term Art Deco only came into popular use in the 1960s, due to a resurgence of interest in the style. The term was consecrated at the Minneapolis Institute of Arts exhibition 'The World of Art Deco' in 1971. Before the 1960s it had no specific title, though was sometimes referred to as 'Style Moderne'.)

Fauré took part in the group display of Limoges enamellers organised by the Limoges Chamber of Commerce. The catalogue states that he showed vases, lamps, and "art enamels", which no doubt referred to enamelled plaques. Other Limoges enamellers exhibiting there included Paul Bonnaud, Jean-Baptiste Issanchou, Léon Jouhaud, Alexandre Marty, Charles Peltant, Jules Sarlandie, and Jeanne Soubourou.

The effervescence and the resolutely 'modern' nature of the exhibition must have been an inspiration for Camille Fauré and his daughter Andrée. There they would have been exposed to the quintessence of the new style, with its blend of diverse influences – cubism, futurism, exoticism – and its celebration of luxurious materials, sleek lines, and intense colours. There was a multitude of inspiration for them at this event, where the exhibitors strove to be 'modern' and to completely break with styles of the past.

The 1930 Salon des Artistes Décorateurs

In 1930 Camille Fauré participated in the Paris Salon des Artistes Décorateurs, this time exhibiting on his own. The Salon was one of the major venues for decorative artists wishing to promote their work. The theme suggested to the participants that year was "the French home, abroad".

Whether or not his pieces were intended to epitomise French style, this is the first detailed description that we have of a public showing of Fauré vases that are clearly in the Art Deco style. The Salon des Artistes Décorateurs catalogue describes his pieces as follows: "A vase decorated with translucent and metallic enamel, created in collaboration with Valade. An enamel vase. A vase decorated with translucent and metallic enamel, created in collaboration with Jacquement."

The two artists cited, Valade and Jacquement, were working in Limoges at the time. Louis Valade, whose work as a designer for Fauré was cut short by his death at an early age, had trained as a porcelain painter. Edmond Jacquement also designed some enamelled plaques for Fauré, in a style similar to that of Léon Jouhaud.

In collaborating with these artists to produce pieces whose juxtaposition of translucent and metallic enamel announces their resolutely 'modern' style, Fauré seemed to be sending the message that his atelier was a contemporary workshop on a par with the finest.

Indeed, many of the Salon's exhibitors were renowned exponents of the Art Deco style: Edgar Brandt, Jean Dunand, René Lalique, and Jacques-Emile Ruhlmann. Among the other

Limoges enamellers present were Jules and Robert Sarlandie, who showed a vase that was produced by the Ateliers Legédé, and Jeanne Soubourou, who showed various enamel objects with silver incrustations, and objects in painted enamel.

The process and the technique

At first, the productions of the Fauré atelier, as well as those of other workshops, tended to be brightly-coloured designs based on floral or animal themes, a development of the Art Nouveau style that had been popular at the end of the 19th century and beginning of the 20th century. When the new Art Deco style developed, the Fauré workshop – like others in Limoges – enthusiastically embraced it, adapting it to vases, lamps, and other objects. A technique featuring thick layers of enamel applied so as to create sculpted relief was particularly suited to the Art Deco style. It was to become a Fauré trademark. It took about two weeks to complete a vase; very large and complex ones could take up to two months to make and the prices reflected the skill and time required to make the pieces. (A rare price list from the 1970s [see table] highlights how a portrait was almost double the price of a landscape or floral piece. These prices are quite high, and show that the clients at the time did not hesitate to spend considerable sums of money on decorative enamelled objects.)

Plaque, based on a drawing by Jacquement

Prices of enamelled plaques in the late 1970s	
T 5 (7.5 x 10 cm / 3 x 4in) Landscape or flowers Portrait	 340 FF 560 FF
T 4 (9 x 14 cm / 3½ x 5½in) Landscape or flowers Portrait	 560 FF 870 FF
T 3 (20 x 25 cm / 8 x 10in) Landscape or flowers Portrait	 1750 FF 3100 FF
T 1 (15 x 20 cm / 6 x 8in) Landscape or flowers Portrait	 1150 FF 1730 FF
T A1 (26 x 35 cm / 10¼ x 14in) Landscape or flowers Portrait	 3150 FF 5400 FF

(From the notes of Mauricette Pinoteau, a former Fauré enameller)

The vase forms, which were furnished by the Edmond Allain firm in Paris, were all given proper names: 'Raymond', 'Lucien', 'Antonio', 'Yvonne', 'Diane', 'Azyadée' and many others. Often these were the names of family members or workshop employees.

Drawings indicating the vase form and a design adapted to it were kept in notebooks, to which the enamellers referred. Sometimes these were rapidly done sketches, while others were far more detailed, being finished with watercolours, with notations giving the names and numbers of the enamel powders to be used.

The designs were given numbers; about 450 of them were created in the years between 1926 and 1930, not including a few designs for ashtrays, boxes and small bowls. Vases were also mounted as lamps, as can be seen in some of the sketches, which may partially account for the popularity of certain of the designs.

The preliminary drawing of the design was transformed into a mathematically measured grid, which was then carefully transferred onto the vase form by means of tracing paper pierced with holes. The dots were then joined using a paintbrush, forming outlines to guide the enameller. This is crucial for Art Deco vases, whose motifs were so carefully balanced that they could not be done freehand. The design, which was perfectly calibrated taking into account the vase's particular shape, had to be rendered with great exactness for the vase to be successful.

The training

There were two categories of new employees hired by Fauré. Some young people, at the age of 16, began an apprenticeship – which lasted two years – being gradually initiated into the art of enamelling. However, Andrée Fauré Malabre liked to have a certain number of skilled workers in the shop, and for these positions, she would recruit people who had studied the Fine Arts.

Once a new enameller had grasped the technique he or she would initially do only floral designs; these were always in demand. Different types of design required varying degrees of skill, some necessitating a complete mastery of enamelling techniques, composition and perspective. It took between two and three years to acquire these skills. Once this was achieved, the enameller generally moved on to geometric designs, or studied the grisaille technique or the reproduction of paintings.

Enamelling is an extremely time-consuming process, and particularly so in the case of the high-relief Art Deco style vases. The fact that the Fauré enamellers never felt rushed or pressured may, paradoxically, be one of the reasons they succeeded in producing such numbers of extraordinary objects, which require the enameller to be given unlimited amounts of undisturbed time.

In the words of Mauricette Pinoteau, a Limoges enameller who was trained at Fauré and worked there between 1972 and 1978: "Le travail de l'artiste est fait de calme et de sérénité, de patience, de générosité. Toutes les qualités sont requises pour mener la matière au degré de perfection."

[The artist's work grows out of calm, peace of mind, patience, and open-heartedness. All these qualities are required to bring his materials to a state of perfection.]

The Fauré style

During the '20s and '30s, the Fauré workshop was buoyed up by the excitement of taking part in the creation of a new style. The graceful, curving floral motifs and swags, the sharp-angled geometric patterns with which the enamellers decorated their vases, perfectly expressed the era. It was a time of new ways of seeing, as embodied by cubism in art, or by the paintings of Sonia Delaunay, with their vibrant colours and pulsating geometrical forms. It was a time of intoxicating technical progress, as expressed by the futurist movement when it proclaimed the "beauty of speed", a time of streamlining, fast automobiles, and the entry of the machine into daily life. The artists of the period felt that their creations were in harmony with the times. The resulting exhilaration is palpable in their work.

There were about a dozen enamellers in the Fauré workshop during the early period of Art Deco production. In later years, though the style was less in harmony with the times, with the workshop producing mostly florally decorated vases and landscapes done after well-known paintings, the emphasis on high-quality work remained a constant.

Even later, the Fauré workshop retained the nurturing atmosphere that had been created by Camille Fauré and maintained by his daughter Andrée. In the words of Mauricette Pinoteau: "They gave us *carte blanche* as long as we produced work in the style and spirit of Fauré. There were no technical constraints or pressure to work fast. It was not an assembly line process; each enameller produced his own

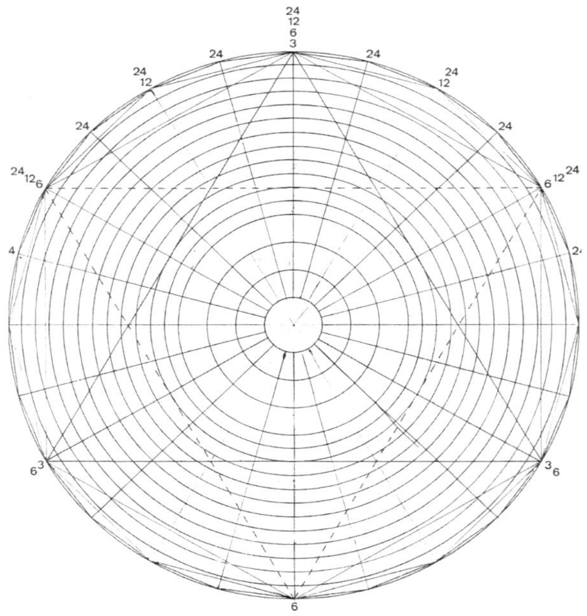

Geometrical grid (called 'poncif' in French). The vase was placed on top of the grid, which was used to trace regular guidelines for the enamel decoration. Such grids were necessary for the highly geometrical Art Deco style vases

piece from start to finish. There was a team spirit, and we all worked together in harmony, with the older people helping and encouraging the younger ones. However, M. Bardy always traced the designs for the Art Deco vases. They did not trust the enamellers to do it themselves, and especially not the women, who according to the prejudices of the time were not considered 'mathematical' enough for the task."

Over the years, Fauré acquired the image of being a "vieille maison" – an old firm, with time-honoured traditions. During the years after World War II, enamel workshops sprang up in Limoges, and at certain times there were as many as 150. Yet the Fauré workshop stood out among all of them. As Paul Buforn, who worked at Fauré from 1970 to 1975, put it, "There is a sort of 'Fauré myth': anyone who experienced it feels a need to speak of it."

All those at Fauré felt the importance of the ideal of constantly striving to do better. While the younger enamellers were encouraged and guided by the older ones, the young people were respected, too, and sometimes were able to contribute new and useful ideas.

Fauré and Art Deco

During the Art Deco period, lacquer and enamel, both ancient techniques newly re-discovered, were much in demand. Both materials present the smooth, shiny finish that was appreciated both by Art Deco designers and a novelty-seeking public.

Enamel lent itself particularly well to the Art Deco aesthetic. A new decorative vocabulary developed, based on the medium's intrinsic qualities: intense colours that could be either richly opaque or translucent and shiny; and the effects of volume, with gently curved floral motifs or sharp geometrical designs. Flowers, spirals, and geometrical constructions were often merely an excuse for unchecked exuberance.

At first the Art Deco movement embraced flower motifs and geometric lines; then new forms were sought out, including geometrical designs in high relief. Rhythm was created not just by the disposition of the design motifs, but also by the opposition of translucent and opaque enamels, and the luminosity of the underlying 'paillon'. The use of metallic enamels such as platinum gave contrast.

Fauré vases featured strong and original avant-garde design, with the juxtaposition of translucent and opaque enamels, with high relief and contrasting colours. The design, whether geometric or organic, sublimated the shape and size of the vase form. Since the aesthetic remained deeply ingrained in the workshop's repertoire, it can be difficult to date all vases with precision.

The influence of cubism is evident and was widely accepted by the public. The workshop launched spectacular decors: colourful and varied. The high relief, as executed with consummate skill by the Fauré workshop, became one of its trademarks. In its ever more daring use of innovative motifs and techniques, the atelier became one of the foremost spokesmen of Art Deco of the day.

World War II and after: separate businesses

During World War II, the enamel workshop remained in activity, while the sign-painting and advertising business all but shut down; a bit of house painting allowed the sign-painting branch to survive.

In 1947, Camille Fauré officially separated the two branches of his double business, entrusting the sign-painting portion of it to his daughter Réjane and her husband Pierre Lavaurs, and passing the enamel workshop over to Andrée and her husband Louis Malabre. This amounted to the official recognition of an existing situation; for a number of years, it had indeed been the Lavaurs family who ran the sign-painting business and the Malabres who directed the enamel workshop.

A document dated 16 October 1947, drawn up by a lawyer, Maître Lachau, records this transaction. It states that the new enamel company would take the name "Atelier d'Art C. Fauré", and that Camille Fauré has contributed "the branch dealing with the creation and production of enamels that is part of an enterprise producing enamels and signs currently exploited in the rue de Tanneries, No. 31..."

While Camille Fauré was recorded as bringing clientele, stock, material and furniture to the new enterprise for a total amount of 400,000 francs, Louis Malabre was stated as contributing tools and stock for a jewellery-making business, also located at the rue des Tanneries address. This, clearly, was the portion of the enamel workshop that was used to make jewellery – to what extent it had truly been the contribution of Louis Malabre is now long forgotten. In any case, each associate brought to the new company a contribution of 400,000 francs, making a total capital of 800,000 francs. The company, it was noted, would be administrated by M. Malabre.[1]

1. Document from the archives of the Etablissement Allain, Paris.

Exhibition vases
Photographed in 1951

75/35 René Primrose Azyadée II Antonio

Marcel 75/25 Emile Vincent Mathias

Atelier d'Art C. Fauré

The workshop, in this new form, would continue to exist until 1985. At the height of its commercial success in the 1950s, there were as many as twenty-five people working at the company, including the enamellers, the furnace man, the secretary, Odette Lajudie, and M. and Mme. Malabre. Although now officially called "Atelier d'Art C. Fauré", this appellation was used rather loosely, and the company was generally called simply "Les Emaux Fauré".

They had begun using an electric furnace after the war, but prior to that, firing had been done in a coke-burning furnace. The furnace man had to be very strong and have a steady hand, for the large vases had to be placed delicately in the oven at arm's length using a long-handled shovel, then very carefully be slid off the shovel without falling or being upset. When the furnace door was opened, one received the full blast of its heat: up to 900 degrees Celsius (1652°F).

An anecdote told by Fauré's great-grandson Jean-Michel Malabre illustrates the great skill required for the firing process. In the 1970s, the French government commissioned the Fauré firm to create an enamel portrait of the Shah of Iran's wife, Farah Diba. Tremendous care had been taken with this portrait, which was to be presented to the Shah as

Raymond Françoise Louis Diane Jeanne Léon Hook

Gérard Jack Célimène Simon '199' Soliflor Moritz Bleuet

No seconds:'oven disaster'. Enamel melts down when fired for too long.
The piece is plunged into acid to remove the enamel and the copper is reused

a birthday present. When the beautifully executed plaque – the work of M. Zussino – was almost finished, a crack was discovered to have formed during one of the firings. This problem, always extremely difficult to repair, was compounded by the fact that the jewels in the Shabanou's diadem were formed of small lumps of enamel that could not be allowed to melt. In addition, the jewels were backed with paillon foil, which can undergo only a certain number of firings and no more. The only hope was that the enamel could be gently persuaded to melt sufficiently to fill in the crack, without overheating the piece. While his colleagues stood anxiously by, M. Faure carefully introduced the plaque into the oven and removed it almost immediately. He patiently repeated this process, again and again, for over an hour until the crack had been repaired, skilfully leaving the 'jewels' intact.

On the days when a firing was to take place, the furnace man arrived half an hour early to light the fire. After about two hours, he would open the door and judge the temperature by the colour of the incandescent furnace walls. When his eye told him that the furnace had reached the correct temperature, he would carefully place each piece to be fired in the oven, leaving it there just the right amount of time. This would continue until all the pieces had been fired. The furnace man could not leave his post on firing days; his lunch was brought to him as he stood there, attentive only to the furnace and the enamel being fused within.

It is said that most of Limoges' best enamellers had learned their trade in the Fauré workshop. Indeed, the Fauré enamellers formed a sort of separate caste among the enamellers of the city. The atmosphere in the workshop, unlike that of some others, was not one of jealousy or competition; the enamellers felt they were all working together toward a common goal of excellence. Frequently, the more experienced artisans would train the younger ones, teaching them their own personal 'secrets', of which each enameller had a certain number. This, however, was not organised, but happened spontaneously, and by affinity.

Many people worked at Fauré their entire lives. No technical or time constraints were imposed on them. Suzanne Coiffe, for example, enamelled slowly, taking her time to create the beautiful, delicate grisaille portraits that were her particular speciality. No one ever tried to make her work more quickly or produce more pieces.

In other enamel workshops, production often took place in an assembly-line fashion. In this sense, the Fauré workshop was an anomaly. Each enameller was in charge of the piece he or she was creating, from start to finish. The emphasis on high quality meant that pieces which did not meet with Fauré's high standards were destroyed. There were never any 'seconds' sold at cheaper prices. If a defective piece could be easily repaired, by re-applying enamel and carefully re-firing it, this was done. However, if the piece was judged irreparable, it was placed in an acid bath to remove all the enamel from the copper, and the entire enamelling process began again.

Some of the enamellers and employees

Odette Lajudie was the secretary at the Fauré Company for almost forty years. She began working at Fauré in 1948 when she was very young, and over the years became one of the most important members of the team.

Monsieur Pierre Bardy began working at Fauré around 1919, and became in a sense the foreman ("chef d'atelier") of the workshop during the '30s and '40s. He was a great master of famous Fauré high reliefs. For the important Art Deco pieces, M. Bardy would usually trace the complex patterns of preparatory outlines onto the vase, this delicate work being entrusted to no-one else.

Lucie Dadat entered the Fauré workshop in 1928 and became one of the best Art Deco style enamellers; she remained there for forty years. Often the application of the enamels was entrusted to Lucie Dadat, who was highly skilled. Lucie Dadat and Pierre Bardy were among the workshop's best Art Deco enamellers.

Monsieur Cheyroux, one of the early enamellers, stayed for many years; he was particularly known for his grisailles.

Baptiste Michaud, called 'Tistou', tended the oven – a difficult task, and a crucial one. Before him, Monsieur Texier had worked at the oven.

Gabriel Faure, who succeeded Michaud as the furnace man, also did framing work on days when no firing was going on. He also sometimes placed the initial coat of fondant on the vases, before they were decorated.

Monsieur Chéron was a goldsmith who also worked on the jeweller's lathe. He prepared the settings for the jewellery, mounting medallions in their brooches to make pins and necklaces and bracelets, and setting and framing the enamel plaques.

Suzanne Coiffe, who began working around 1946, was taught by M. Cheyroux and, like him, became a specialist

1940s. Workers at the Fauré workshop. 1. Monsieur Chéron, 2. Louis Malabre (Fauré's son-in-law), 3. Baptiste 'Tistou' Michaud, 4. Suzanne Coiffe, 5. Andrée Malabre (Fauré's daughter, Louis Malabre's wife), 6. Pierre Bardy, 7. Monsieur Reyrat, 8. Lucie Dadat, 9. Monsieur Cheyroux, 10. Gabriel Faure

1960s. Part of the Fauré team. From left to right: Monsieur Gabriel Faure; Monsieur Rozier; Madame Suzanne Coiffe, enameller; Madame Lucie Dadat, enameller; Madame Vedrenne, paillonage-silver-foil; Mademoiselle Jarry, enameller; Odette Lajudie, secretary; Andrée Malabre (Fauré's daughter), owner of the atelier; and Gérard Malabre (Fauré's grandson, Andrée's son), manager of the atelier

The 3 'aces' at Fauré. From right to left: Madame Lucie Dadat, Monsieur Pierre Bardy, Madame Suzanne Coiffe and an unidentified worker in the background.

of grisaille work. Her husband, Michel, had been with the company since he was 15, working in the sign-painting department since his return from the Second World War.

Mademoiselle Jarry, who began working at Fauré in around 1935, specialised in landscapes after Corot, which she rendered with a particular talent.

Madame Vedrenne was an enameller who often did floral vases and some Art Deco pieces.

Jean Dutheil specialised in grisaille work.

Raoul Martineau, one of the earlier enamellers, was, apparently, a difficult man to get along with. Overbearing and aggressive, he angered easily and used to threaten his colleagues with lawsuits: "We shall see each other in court, Sir". Despite the anachronism, he prided himself on dressing and behaving like an "aristocrat", wearing gaiters and carrying a small whip. He apparently had a soft spot, though, for he kept chocolate in his drawer for visiting children.

Martineau was considered so unpleasant that in the end he was given his own isolated cubicle, where he enamelled all alone. Yet he was tolerated in the Fauré establishment because of the superb quality of his work, which often consisted of "genre" scenes.

Monsieur Rozier, who was a specialist in the Art Deco techniques, came to Fauré shortly after the war, around 1948. He had his own special secret for producing the subtle graduations of colours so characteristic of the Art Deco style.

Joseph Zussino, an enameller, who worked for many years in the workshop. His daughters Christiane and Chantal also came to work at Fauré during the 1970s, when each was about 16, staying for at least a decade.

The later generation of enamellers included: Mauricette Pinoteau (a 'pupil' of M. Rozier), Paul Buforn, Joëlle Comes, Danielle Chouzenoux, Hazard Laguionie, Simone Lacorre, Annie Soumagnat, Georges Ferdel, Lo Van Chay, Dominique Gilbert, Brigitte Gorce, Jean-Marie Goudet (who specialised in the 'surimpressions' invented by Gérard Malabre), and M. Langer (who did the enamelling for the costume jewellery).

Camille Fauré, c.1950

The gravestone at the Louyat cemetery in Limoges for the families Fauré, Lajoie (the family of Camille's sister, Madeleine) and Lavaurs (the family of Camille's daughter, Réjane)

The last years of Camille Fauré

Madeleine, a soft-spoken woman and devoted wife to Camille, much beloved by everyone, died in a car accident on 29 May, 1950, at the age of 67. Camille died just six years later, on 7 October 1956; he was 82. He had fallen from a ladder while trying to take down some shutters at his country home 'Le Surzol', breaking his femur. It had never healed very well, and he had felt poorly for many months. Camille Fauré died in his bed in the rue des Tanneries apartment, just above the enamel workshop. He is buried in Limoges' Louyat cemetery.

Today Camille Fauré is honoured by having a street in Limoges named after him. (The date of birth shown on the street sign is incorrect; he was, in fact, born in 1874.)

The Malabre era

Perhaps the most commercially successful period for the Atelier d'Art Fauré began in the late 1940s, when the workshop was run by the Malabres. While Andrée Fauré Malabre ran the workshop, ensuring the continuation of the high quality that had made the company's reputation, Louis Malabre created and maintained their commercial outlets.

Andrée had been a frequent visitor to the workshop from a young age. From 1922 to 1924 she had studied at the Limoges School of Decorative Arts. She had been an excellent student (Henriette Marty, two years her senior and also an outstanding student, had preceded her at the school). The curriculum included anatomy, perspective, ornamental composition, live plants, figure drawing, embroidery, ceramic painting, and flower painting. Among Andrée Fauré's teachers, the painter Charles Bichet (1863-1929), who taught drawing, is said to have influenced her vocation.

After finishing school, she went directly to work in the Fauré workshop, assisting her father. Andrée gradually became the centre around which her father's enamelling workshop revolved, and Camille came to rely on her. An ever-present figure in the workshop, she had her own special place at the enamellers' bench. She would sit there and help with the application of gold or silver paillons if there was a rush order.

Andrée kept an eye on everything, and she liked things to run smoothly. Genuinely interested in the lives of all her employees, she hated conflict and always did her best to preserve a happy working atmosphere. Even today, she is remembered by the enamellers with great affection, as a warm and generous woman who was more like a grandmother than an employer. Nevertheless, although she did not enamel herself, she had a deep knowledge of the technique, and never failed to notice even the most minor defects. She would call these to the attention of the enameller who had produced the piece, tactfully but firmly; Fauré would not tolerate below-standard work.

While Andrée tended the workshop, her husband Louis Malabre promoted the enterprise, seeking out new clients and maintaining constant contact with his network of retailers (for example the shop "Au Vase Etrusque" in Paris, which made a specialty of Art Deco vases). Louis Malabre structured Fauré's commercial network with an infallible commercial talent.

Constantly on the lookout for new clients, in the 1950s Louis attempted to develop the North African market. Twice he loaded up the family car with enamelled pieces, and drove through several countries, including Morocco, Tunisia, and Algeria, where he presumably sold many pieces.

The clients

About 80% of the production went toward filling orders, while the remaining 20% was sold at the shop in Limoges. One tenth of the shop's customers were foreign buyers.

The South American market was one of the earliest and most active of the overseas markets. Wholesale buyers of porcelain goods would also stop at the Fauré workshop and purchase the latest models, which they would include in their shipment of porcelain goods. The North American market later followed suit.

At the height of Fauré's commercial activity, the firm had clients in all the major French cities, as well as in Japan, the Americas, Switzerland, Italy, and Australia, which represented a considerable market.

An entry in the Edmond Allain account books for July 1951, records that the Malabres had bought the following copper forms:

20 Diane vases	485 FF
32 Célimène	345 FF
6 Suzanne	1090 FF
1 Arthur	635 FF
7 Hook	380 FF
14 Boule 1931 (14 cm / 5½in)	525 FF
20 Jack	290 FF
56 199	290 FF
50 Monitz	270 FF
18 Solange	1090 FF

The entry further states: "Vases taken away by Monsieur Malabre on the roof of his car; Madame Malabre was there, as well as their son."[2]

This order was probably especially large because the Malabres had come to transport the copper forms themselves and therefore would not have to worry about shipping. Normally, their orders from Allain were smaller, and the quantity purchased often varied widely from month to month, depending on the orders received and the number of pieces sold.

Among Fauré's usual clients, retailers who regularly bought Fauré pieces for resale were: Au Vase Etrusque, 20 Boulevard Malesherbes, Paris; Grand Dépôt, 21 rue Drouot, Paris and 33 rue St. Ferréol, Marseille; Maire, rue St. Ferréol, Marseille (who bought mostly painted plaques); Armand Blisson, 16 rue Pavillon, Marseille (who bought mostly vases); and Maison Moulard, rue Gambetta, Lyon.

2. Account books, archives of the Etablissement Allain, Paris

During the 1950s, one of Fauré's advertising slogans put the accent on the international nature of the clientele. It proclaimed: "Emaux Fauré – La marque mondiale" [Fauré enamel – the international brand].

Retailers of Fauré enamels. From left to right: Au vase Etrusque, Paris; Armand Blisson, Marseille; Grand Dépôt, Paris and Marseille; Maire, Marseille

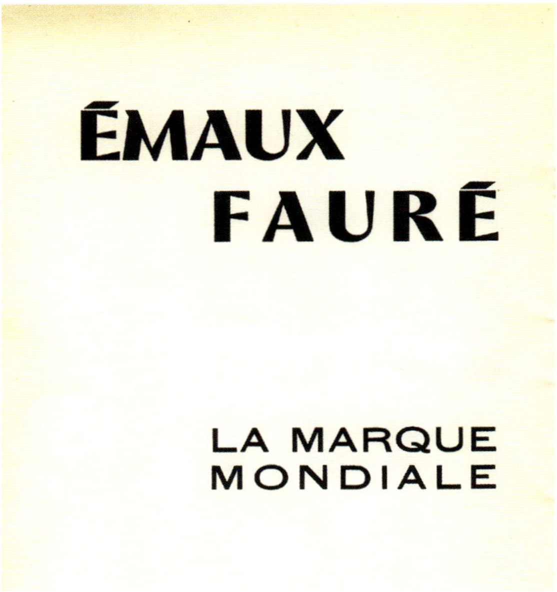

Advertisement from the 1950s. An early incursion into globalisation

Advertisement from the 2e Salon Internationale des Industries et Art du Feu, in Paris

Probably the last advertising venture for the Fauré workshops, which is still found today in some Limoges hotels. 1985

The Fauré's country house, 'Le Surzol' in 1925

'Le Surzol' as it stands today

'Le Surzol'

In 1925, Camille Fauré bought 'Le Surzol', an old house and farm with 16 hectares of surrounding land, on the outskirts of Limoges. He had just finished a big job that involved painting gasoline pumps throughout France for the Azur gas company. It had been a long and difficult job that had involved a lot of travelling, particularly for his son-in-law Pierre Lavaurs, but it was the money earned from this job that enabled him to make the purchase. 'Le Surzol' was to become the centre of the Fauré family's social life, and probably the true centre of Camille Fauré's life from that time onwards.

A majestic alley of beech trees leads to the house, which may date back to the early years of the 19th century, while its basement, with a vaulted ceiling, may well have been constructed during the 16th century. The house, with five bedrooms, is not overly large. Consisting of two floors, and a round tower in its middle section that contains the staircase. The living room looks out onto a sweeping vista of grass and trees. Near the house is a small tower that used to house farm animals, and which was transformed by Fauré into sleeping quarters for the maids. In its full glory, 'Le Surzol' was a working farm, with fifteen cows in a stable facing the house, and servants who took care of the property and ran the farm. Even today, 'Le Surzol' remains the family's country house, where everyone gets together on special occasions.

It was in the country that Camille Fauré was the happiest. He organised many hunting parties at 'Le Surzol', and the children always loved going there. The family spent as much time there as possible, especially during the holidays, and it was the setting for many happy family gatherings. Once Robert Lavaurs had obtained his driving licence, Camille would often ask his grandson to take him to 'Le Surzol' – a short 15 or 20 minute journey – for though he was able to,

A gate within the grounds of 'Le Surzol', bearing the date 1896

Camille Fauré and his wife Madeleine at the entrance to the Surzol driveway, c.1950

'Le Surzol' painted by Pierre Lavaurs

Camille Fauré never drove himself; he much preferred to be driven. In the face of his grandfather's insistence, Robert would have very little choice in the matter. He would have to drop whatever else he had been planning to do, and take his grandfather out to the country.

During Camille Fauré's lifetime, 'Le Surzol' was a functioning farm that brought in revenue. As soon as he arrived, Camille would go speak to the caretakers, requesting the latest news, asking how the animals were, which ones had been sold and for what price, inquiring about the ones that were sick, about a cow that was about to calf, about when the machines were coming to harvest the hay; in short, all the details involved in the daily running of a farm. A man of numbers, Fauré would spend many hours poring happily over the farm's accounts.

Family gathering at 'Le Surzol'. Seated, from left to right: Blanche (Fauré's sister), Louis Malabre with Georges (Réjane's son) on his lap, Andrée Malabre (Fauré's youngest daughter), Pierre Lavaurs, Réjane Lavaurs (Fauré's eldest daughter), Madeleine (Fauré's wife) and Camille Fauré (wearing white shoes). In the foreground, Réjane's son Robert plays with the family dog

Enamel detail of buffet by Delage

The living room of 'Le Surzol' has a picture window that looks out onto a terrace where the family used to gather, and still does, to this day. There, Camille Fauré would sit with his wife, his brother-in-law Léonard Lajoie, and his sister Madeleine. On weekends, his other sister Blanche would often come from Paris, bringing with her chocolates from her shop. Camille's daughters and sons-in-law and their children would also be there, the children playing and the dogs – of which there were always several – frolicking.

Inside the living room, standing against the wall opposite the picture window and taking up nearly the width of the room, is an imposing wooden buffet. It was designed and built by cabinetmaker Delage, who had studied at the Ecole Boulle, the famous Parisian school of furniture design, and features ornamental enamel plaques from the Fauré workshop in its upper corners. This was no doubt either a special commission by Fauré or an experimental piece; no others like it are known to exist.

On the living room walls, alongside sketches done by a young Gérard Malabre, are the portraits of Camille Fauré's favourite cows, Cara and Mélou. There are no other photographs on the walls, and only one other, of Camille and his wife Madeleine, sits on a side table. Displayed prominently in frames on two other walls are the prize diplomas given to Fauré by the Horticultural Society of Haute-Vienne, one awarding him a silver medal and the other a gold medal.

Robert Lavaurs was very close to his grandfather. At 'Le Surzol' they went on memorable fishing expeditions, setting out for a stream one kilometre (over half a mile) from the house. Only the men went, says Lavaurs, the women stayed home. When they arrived, they set out their traps and waited. For Robert, then a little boy, the twenty-minute wait seemed interminable. He would ask: "Is it time yet, Pépé?" and Camille Fauré would answer: "Not yet. It'll be time when I've rolled a cigarette and smoked it". He would proceed to roll himself a leisurely smoke, under the impatient eyes of his grandson.

When they opened the traps, they usually found they had captured about 100 crayfish. They carted them back home, where the table was set for a late dinner. The maids prepared a feast of potatoes and crayfish, boiling them in huge pots, and the family would si- down to eat around 9:30 that night.

The lateness of the hour meant the maids would have stayed up past their usual bedtime. To thank them, two or three days later the family would drive them to the neighbouring village of St. Léonard, for the village dance.

Buffet by Delage, enamel details by Fauré. 'Le Surzol', Limoges

Fauré showroom, 1970s. In the foreground Gérard Malabre poses next to an enamel plaque (see page 245). His mother, Andrée Fauré Malabre is seated in the background

Gérard Malabre

Louis Malabre died around 1965, leaving his wife Andrée to run the workshop alone. Although their son Gérard was a pharmacist, after his father's death he became much more active in the firm, helping his mother run the workshop. Gérard had the bohemian temperament of an artist, and could be generous to a fault. Along with these qualities, he also had a certain nonchalance that is more fitting for an artist than a businessman.

Having grown up in the milieu of enamelling in Limoges, Gérard Malabre was impatient with what he considered the sluggishness of the craft. He wanted to revolutionise enamelling in Limoges, to break with certain weary traditions, and to restore beauty and originality to the craft.

In seeking out new techniques, Gérard invented a technique he called "surimpression". This involved the superimposition of a grisaille portrait, as if in a black and white photograph, over a coloured background such as a landscape.

During the 1970s he also revived the idea of enamelling with metals. This technique – which involved using a liquid metal, such as platinum, along with the enamels – was common during the Art Deco period. The metals were juxtaposed with opaque enamel and were not heated until complete fusion, resulting in a cracked, granulated texture. Gérard also launched a line of jewellery, with varying degrees of success.

One of the notable commissions during this period was a series of enamelled plaques for Air France, which the Company placed in their fleet of planes in homage to enamelling as one of the time-honoured traditional arts of France.

From 1971, enamel exhibitions began to be organised in Limoges every two years. The first President of this event, called the Biennale, was the enameller Georges Magadoux. Gérard Malabre, who became its President from 1982 to 1988, wanted to use the Biennale to promote enamelling. He hoped to make it a truly international gathering, where contemporary artists could meet, exchange ideas and see each other's work, and where the public could learn of the latest developments in contemporary enamelling. Malabre felt that the repertoire of Limoges enamellers had become too limited, encompassing only copies of famous paintings, or the classic bouquet of flowers.

As a result, under his direction the Biennale embarked on several international projects, inviting enamel artists from Moscow, St. Petersburg, and Riga, in the then Soviet Union. The project succeeded, which was surprising for the period, when the Soviet Union was only just beginning to feel the effects of *perestroika*. The encounter produced a veritable culture shock on both sides.

Gérard Malabre initially went to the Soviet Union to meet the enamellers. He was impressed by the extraordinary people and artists he met, and by their determination to continue with their creative work, despite the absence of a market. Following that initiative, an international enamelling workshop was organised in Hungary. The Biennale's office also managed to bring enamellers out of China, though this proved more difficult.

In the end, Gérard Malabre and his team, including Danielle Soury McCann, were able to bring enamellers from forty different nations to the Biennale.

Innovative designs of the Fauré workshop from the 1970s

A novel design created by Gérard Malabre using an enamel and photographic technique

Mixed technique panel created by the Fauré workshop for one of the Limoges enamel Biennales

Simone Lacorre, 1975

The final years

During the 1970s there were between twelve and fifteen enamellers working at Fauré. Along with the secretary and salesperson and others, this made a total of about twenty people.

By this time the Fauré workshop had long since stopped producing Art Deco pieces, which had by then gone out of fashion. Their production was mostly limited to florally decorated vases, or to plaques with tapestry themes, woodland landscapes, or reproductions of well-known paintings. These plaques were often placed in black frames, which was what the clients were used to and generally requested, with no consideration of whether the frame complemented the enamel scene or not.

A portion of the Fauré clientele, particularly from the 1960s to the 1970s, were tourists who came in by the busload, and more often than not walked out with a souvenir enamel. In the mid-1970s, however, this began to change. Fewer tourists came, and those who did had a lower budget and did not necessarily want to buy enamel. In addition, the market was changing, and commercial networks were becoming increasingly international. It was more and more difficult for a small local studio to survive.

During this period the company began to encounter its first serious financial difficulties. Other problems came to compound that of the dwindling clientele: the oil crises that shook the world's economies, increasing social charges in France, and, to make matters worse, an entire container full of enamelled pieces disappeared while on the way to its destination somewhere in France. The goods were never received, and Fauré was never paid, resulting in a heavy loss for the company. Andrée Malabre, in particular, was very upset over it.

Gérard Malabre, knowing that the enamel workshop was his mother's whole life, did his best to maintain the enterprise for as long as she lived, though he was not a manager by nature. Very shortly after her death, on 30 March 1985, he was unable to pay the company's debts and was forced to declare bankruptcy. A judiciary administrator was appointed, and the liquidation of the company took place on 12 April 1985. Though he was not given much choice in the outcome, Gérard hoped to at least maintain the jobs of his employees and expressed the wish that this be taken into consideration when choosing between the possible purchasers.

This period was a very difficult one for the family, and especially for Gérard Malabre, who had to witness the disintegration of the company his grandfather and parents had created and made their life's work.

The Turunen era

The business was finally sold to a Finnish woman named Pirkko Turunen, who was not in the enamel business. She was aware of the aura attached to the Fauré name and perhaps believed the workshop would virtually run itself. While she acquired the right to use the name "Fauré" – her company was called the "Société Nouvelle des Emaux Fauré" – she could not use the name "C. Fauré".

Mrs. Turunen purchased the remaining stock and promised to keep the personnel. However, her business did not do well. This was due, no doubt, to many factors, and first of all, to the hard times that enamelling in general had fallen on. The situation was compounded by the fact that Mrs Turunen was a stranger to the business and unfamiliar with the Fauré tradition. The high standards that had formerly distinguished the company's work came to be neglected, with quantity of production being preferred to quality.

During the brief existence of the Société Nouvelle des Emaux Fauré, the company produced vases, plaques, and jewellery, some imitating the old Art Deco styles and others in the more classic floral style. All, however, lack the essential Fauré precision and attention to detail, featuring brash colours and awkward designs. These pieces, graceless as they may be, must still be mentioned as part of the Fauré legend.

Soon, Mrs Turunen was forced to declare bankruptcy. The liquidation of the Société Nouvelle des Emaux Fauré began in September 1987, while the bankruptcy proceedings of Madame Turunen herself took place on 7 December 1987.

Several of the employees had been fired already, among them the long-time secretary Odette Lajudie. The others went their separate ways after the dissolution of the workshop. After Mrs Turunen's bankruptcy, a second sale took place of what little remained, but did not result in the creation of a viable workshop.

Opposite, above: The window displays of the 'Société Nouvelle des Emaux Fauré' under the direction of Pirkko Turunen, c.1986

Opposite, below: The Fauré showroom, 1985

EMAUX de LIMOGES
TOTAL

The Espace Camille Fauré

Following Pirkko Turunen's bankruptcy, Gérard Malabre made one last attempt to maintain the Fauré name, opening the Espace Camille Fauré. This was a gallery of contemporary enamel, where the artists who had exhibited at the Biennale could sell the pieces that had remained unsold, or show new ones. It was, unfortunately, not a commercial success, perhaps due to the size of the city and the limited number of visitors. While people flocked to Limoges during an event like the Biennale, once it was over, things returned to their former sleepy pace. A gallery that depended on daily or weekly visitors was bound to encounter difficulties. Gérard eventually closed the Espace Camille Fauré and sold the premises. He died on 17 July 1997.

After the liquidation of the Atelier Fauré, vases were occasionally sold on a limited basis to a gallery in Paris. Although these are sometimes referred to as "fakes", technically they are not. They were produced by former Fauré artisans, with Fauré materials. They might be considered late, but legitimate, editions of old designs, mostly the Art Deco ones. They are signed "C. Fauré Limoges", but they lack the lustre, detail and look of the old vases. Problems arise only if these vases are improperly represented as having been made in the 1930s. This 'late' production eventually ended in 1996.

The framing department, 1985

The former Fauré premises as they stand today. The old showroom and workshop are now retail stores

The Fauré legacy

Paradoxically, it was during the 1970s, when the Fauré/Malabre enterprise was fighting to stay afloat, that Art Deco enamels – and particularly those of Fauré – began to make a come-back. The Art Deco style was brought to the notice of the art world, and the public, by an influential exhibition held in 1971 at the Minneapolis Institute of Arts. Several Fauré vases were included in the exhibition, and their beauty and originality – as well as the fact that artists such as Roy Lichtenstein and Andy Warhol were discovered to be collectors – attracted attention.

During the 1960s, the New York gallery owners Ileana and Michael Sonnabend had contacted the Fauré workshop and subsequently purchased a considerable number of vases from the Art Deco period. These vases, along with a collection of many other Art Deco pieces, were shown by Mrs Sonnabend, first in her New York gallery in the early 1970s, and later, in the Baltimore Museum of Fine Arts. The exhibition, according to Rita Reif of the *New York Times*, included "stunning examples of Camille Faure's shimmering and frosted enamelled vases, jewel-colored vessels awash with patch-works of Cubist-style geometrics, abstracts and florals".[3]

The vases, which Reif called "spectacular", filled five showcases. Since then, the appreciation of collectors has continued to grow, and Fauré enamels are now in demand all over the world.

After 1996 some Fauré enamellers were approached to produce "new" Faurés. They have refused, feeling that to do so would be to betray the trust that the Fauré atelier bestowed on them, and to make improper use of the skills that they were taught there.

A half-century has gone by since Camille Fauré's death, and it is many years since his son-in-law Louis Malabre and his daughter Andrée, zealous guardians of the Fauré legend, passed away. It is almost twenty years since a much-diminished Fauré enamel workshop, only a shadow of its former glory, closed its doors for good. Yet today, Fauré enamels are more sought-after than ever. They are prized for their beauty and their audacity, for the superb quality of their workmanship and their quintessential embodiment of the Art Deco aesthetic. At the heart of their unique vibrancy, their bold harmonies of colour and form, their generous volumes and dancing rhythms, lies the essence of the Fauré style: the geometry of joy.

3. 'Antiques; Art Deco Works to Relish', by Rita Reif, *New York Times*, 24 March 1985

Enamel: the techniques

Enamel powders. These are mixed with purified water into a paste and then applied with a spatula onto the prepared copper surface

Enamel is glass that has been crushed to a very fine powder and heated until it reaches the point of fusion and becomes hard and shiny. The powdered glass is purified to eliminate any foreign substances that would affect its final texture and colour. The colours are created by the addition of various oxides to the glass powder.

The following colours are obtained from the following oxides:

Blue and turquoise: cobalt and copper
Violet: manganese and cobalt
White: tin
Yellow: uranium and iron
Black: iron
Red and pink: copper and gold.

Red and pink are particularly precious due to their gold content. They are also rather unstable because the oxide does not attach well to the copper base, and therefore require several firings. Black requires several layers to produce a rich, strong colour.

Enamel powders

The enamel powder is mixed with purified water until it forms a paste. This paste is applied to the copper base with a spatula. In the case of grisaille enamels, the enamel paste is applied to the base, and then delicately drawn out with a pin.

The support is usually copper, but it can also be gold or silver, though these are used mainly for jewellery. The Fauré vases were produced using copper sheets with a thickness of 0.5 or 0.6mm ($\frac{1}{50}$ in or $\frac{1}{40}$ in) in the case of the smaller vases, and 0.8mm ($\frac{1}{30}$ in) in the case of the larger ones.

The very first coating of enamel that is applied to the prepared copper form or plaque is called flux, or 'fondant'. Once fired it is transparent. It protects the copper base, and must be evenly applied to both sides, otherwise during the firing process the copper might become distorted, causing the enamel to crack.

Once the fondant has been fired, thin sheets of metal foil, usually silver, but occasionally gold, may be applied. Known as the 'paillon', this foil heightens the natural sheen of the enamel and gives it depth. 'Paillon' foil was often employed for Art Deco enamels. The foil is crumpled and pierced ever so lightly with a brush, to allow heat that might be trapped

The surface of the copper is curved so that the enamel attaches to the metal much better due to its expansion and contraction when heated and cooled. Once the copper plate has been cleaned and polished, a first layer of enamel, the 'fondant', is applied to its back and front. This counterbalances the expansion of the enamel when it is fired

Silver and gold leaf. Micro thin layers of metal 'paillon' are applied to the copper form once it has been fired with the 'fondant', the first layer of enamel. Gold was rarely used due to its high cost

beneath it to escape during firing. Once this is done, the vase is ready to be decorated.

Fire is the catalyst that transforms the enamel powder into a hard, shiny and lustrous matter. This process is irreversible, except if the piece is immersed in an acid bath, which removes all the enamel.

Since copper melts at 1084 degrees Celsius (1983°F), all firings must be below this temperature. Enamel begins to vitrify at 600°C (1112°F). At this temperature, considered low, the enamel's finish is opaque and its texture granular. At 900°C (1652°F), enamel takes on the shiny vitreous look for which it is most often known. However, since each colour vitrifies at a different temperature due to its particular composition, one must begin with the colour that requires the highest temperature and then continue with colours that are fired at diminishing temperatures. Each time a layer of enamel is applied the piece must be fired. The enamel powder, applied wet, must be perfectly dry before the piece is fired.

Different stages of 'preparing' a copper surface before the decoration is applied. From right to left: a copper vase cleaned and polished, a vase with initial layer of 'fondant', and a vase with applied silver foil 'paillon'

Cristallerie de St. Paul near Limoges

The fine art of enamel application

The thick reliefs so characteristic of the Fauré workshop and the Art Deco style are particularly difficult to achieve. They require a thicker granulation of the enamel, an extremely steady hand, and a great deal of experience and skill. Instead of being applied and spread out like paint, they are shaped and sculpted on the copper form.

In the words of Mauricette Pinoteau, the Limoges enameller who perhaps best embodies the Fauré style today: "Enamel is anguish. It allows no mistakes. Its spirit and matter always have the last word."

The Cristallerie St. Paul sign

Spatulas used to apply enamel

Enamelled samples on a plain copper plate

Oven ready to receive enamelled pieces

Soyer enamels, the enamel supplier used by Fauré, is still in existence today.

Part of a sample of enamels offered by Soyer enamels

Edmond Allain
Copper vase forms

The Fauré workshop purchased its raw materials from Paris. The enamels came from the Maison Soyer (then in Paris, now near Limoges), while the copper forms for the vases were produced by the Edmond Allain workshop.

Still in existence today, the Allain workshop, which celebrated its 100th birthday in April 2006, was founded in 1906 at 14, rue Oberkampf by two brothers-in-law, Jules Noury and Edmond Allain. That same year, Allain bought the Petchinouck Company, a metal spinning and embossing firm, that also made vase forms. At the height of the business, there were about 50 polishers, spinners, embossers, and goldsmiths.

Jules Noury was married to Allain's older sister Gabrielle, and was several years older than Allain. Noury had three sons, each of whom eventually founded his own firm; Allain had four daughters (Alice, Huguette, Suzanne and Lucienne), all of whom worked or were involved in the Noury-Allain establishment. Allain's oldest daughter, Alice, was the most involved in the company, working there from 1914 until her death in 1996. In about 1928, Noury and Allain separated amicably.

Bernard Vergne, the current owner of the firm, is the grandson of Edmond Allain; his mother, Suzanne, was Allain's daughter. Monsieur Vergne, who began working in the company in 1966, maintains: "The job of a 'repousseur' is dangerous, manual and dirty. Not everyone likes it nor adapts to it and those that do many times abandon it".

Allain sold copper forms to other Art Deco enamellers besides Fauré, among them Feuillâtre, Sarlandie, Bonnaud, and Marty. Models carried proper names. The Allain archives designate several models as being used by Fauré Marty, among them: Réjane, Love, Hetty, Camille, Lingtan, as well as Bonbonnière.

In the early years, Camille Fauré purchased vase forms that Allain had in stock. Later, new forms were designed, some by Andrée Malabre. During the period when vases were most popular, Fauré had a repertoire of about 50 different shapes. Allain seems to have been Fauré's only supplier of copper vase forms.

Each form was initially sculpted in boxwood and then cut into several pieces, like puzzle pieces, so as to be easily assembled and disassembled. Using a method called spinning, the copper sheet was shaped around these wooden forms, by means of a lathe.

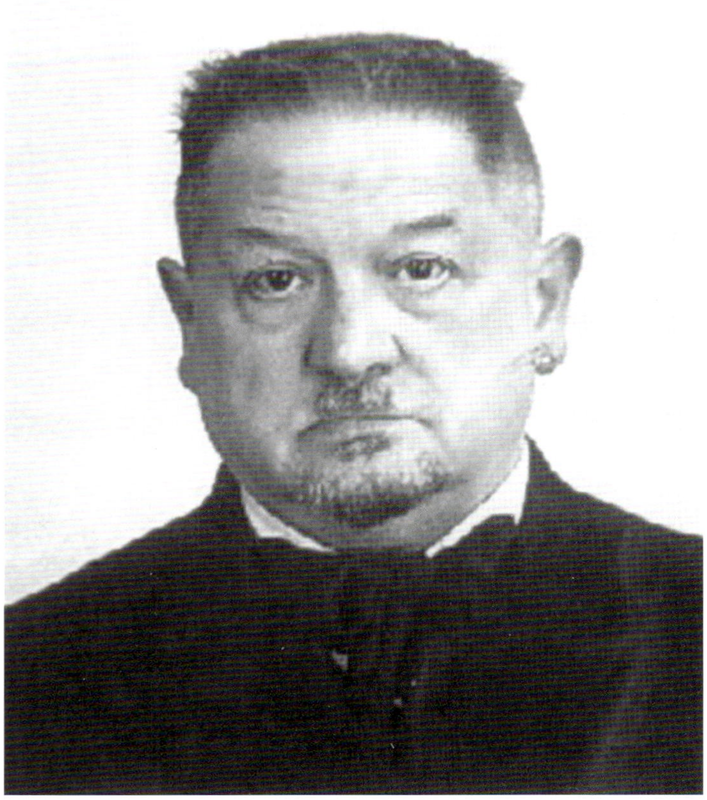

Monsieur Edmond Allain, 1952

Letterhead for the Edmond Allain workshop, 1960s

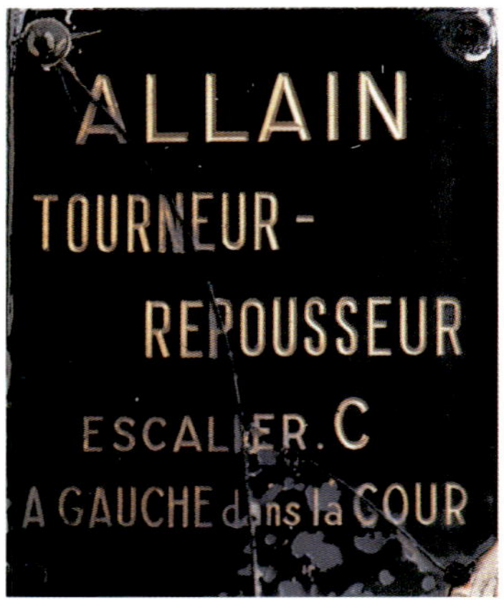

Signs at the Edmond Allain workshop in Paris

Inside the Jules Noury workshop (later to become the Edmond Allain workshop), c.1890

Workmen at the Noury & Allain workshop, including Jules Noury (wearing a cap) and Edmond Allain. In the foreground are young apprentices. The image was taken in 1906, the year that Noury and Allain became partners

Each form was initially sculpted in boxwood and then cut into several pieces, like puzzle pieces, so as to be easily assembled and disassembled

Brass rings for the rims of the vases

The copper used was usually of a thickness of 8 'dizièmes' or 0.8mm. (1/30 in.). Though for the smaller Fauré vases, they usually used 0.5mm. or 0.6mm. (1/50 or 1/40 in.). The copper must be of uniform thickness, otherwise the enamel is liable to crack.

The wooden forms (known as 'outillages') were for the most part made in the 1930s. Around 1950 the Fauré workshop ordered a few others. In 1971, Andrée Malabre ordered several different types of vases.

Fauré sometimes specified "cuivre email" [copper for enamelling], an alloy containing 1% zinc, which is easier to work with.

The bottoms and collars of the vases, added during the finishing process, were treated in various ways to give them different colours: "silvered","old silvered", or "oxidised".

Vase models from the Edmond Allain workshop

A partial list of vase models created by the Edmond Allain workshop for Fauré (dates of creation given when known): Amélie, Amazone (1928), Antonio (1931), Arlette (1954), Arthur (1946), Azyadée (1928), Bruges (1928), Célimène (1929), Cita (1929), Diane (1929), Emile, Fantasio, a.k.a Fausta (1930), Fantasio inverted, a.k.a. Fausta inverted (1930), Gérard (1936), Gladys (1930), Henri (1954), Hook (1936), Jack, Jeanne (1936), Léon (1936), Lisette, a.k.a. Alice (1955), Louis (1929), Lucie (1936), Lucien (1928), Marcel, Marie Paule, Martine (1954), Lucie (1936), Mathias (1973), Monitz, Nicolas (1953), Odette (1954), Oeillet, Orsay (1929), Paul, Pierre (1954), Poire, Primrose (1928), Raoul, Raymond, René (1929), Simon, Solange (1951), Suzanne (1951), Suzon, Vincent.

Brass bases

Solidified sealing wax at the bottom of a finished vase

Copper vase models (dates of creation given where known)

From left to right: 75/30, 75/25, Antonio (1931), Mathias (1973), unidentified

From left to right: Raymond, Françoise (1931), Diane (1929), Célimène (1929)

From left to right: Louis (1929), '188', Hook (1936), Simon, '199', Soliflor, Bleuet, Salambo

From left to right: Gérard (1936), Bonbonnière (pre-1952), Françoise (1931), Jeanne (1936), Jack

From left to right: 75/35, Azyadée I (1928), Paul

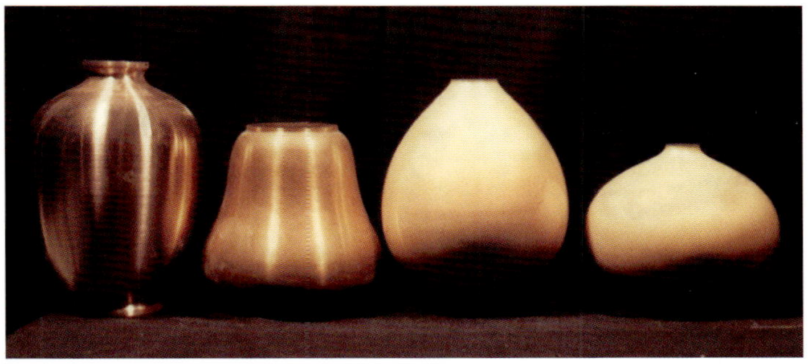

From left to right: Cita (1929), unidentified, Marcel, Vincent

From left to right: Amélie, Lucien (1928), Primrose (1928), Azyadée II (1928)

Aluminum prototypes for Fauré vases

(N.B. Aluminum cannot be enamelled as the enamel does not adhere to it)

From left to right: Hook (1936), Diane (1929), Louis (1929)

From left to right: Arthur (1946), Marie Paule, Suzanne (1951), Solange (1951), 75/30

Amélie

Antonio (1931)

Arthur (1946)

Azyadée I (1928)

Azyadée II or 'inverted' (1928)

Célimène (1929)

Cita (1929)

Diane (1929)

Fausta, also known as Fantasio (1930)

Fausta 'inverted', also known as Fantasio 'inverted' (1930)

Gladys (1930)

Hook (1936)

Isidora

Jack

Léon (1936)

Lisette (1955)

Lucien (1928)

Marcel

Marie Paule

Mathias (1973)

Monitz

Orsay (1929)

Primrose (1928)

Raymond

René (1929)

Vincent

Yvonne

Author in the Edmond Allain workshop in Paris holding a finished Fauré 'Léon' vase (1936), see page 91, and an Airbus jet engine part (1990s) with an aluminium prototype for a Fausta 'inverted' vase (1930). All three items are from the same workshop. The vases and the jet engine part are separated by 60 years

Fauré final approval stamp for the designs of the copper vases

Copper vase made by Edmond Allain and decorated by other artists using acids rather than enamels to achieve different effects

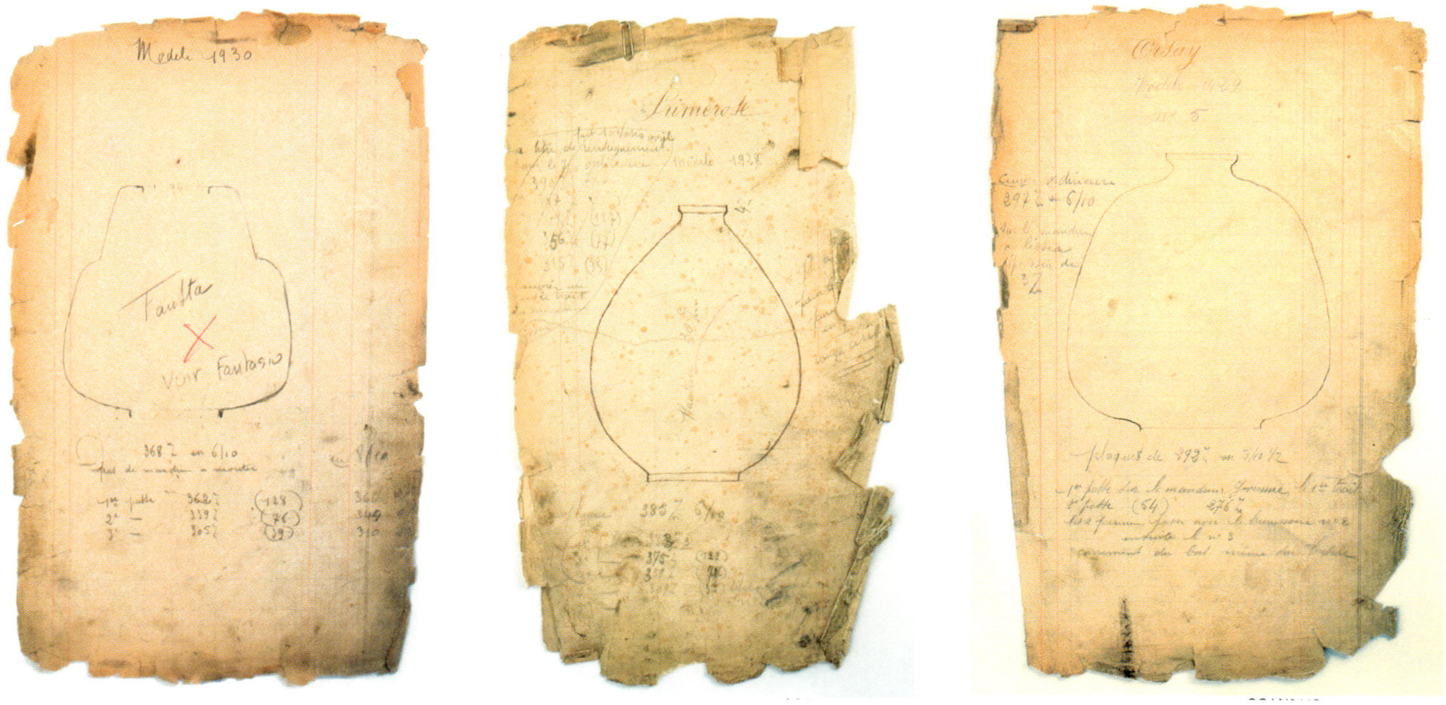

Three designs with precise data for the manufacture of the copper forms: Fausta 'inverted' (1930), Primrose (1928) and Orsay (1929)

Inside the Allain workshop today, with a Fausta 'inverted' vase in the foreground

Warehouse with the old wooden forms known as the 'outillages'

Pattern designs

Pencil and gouache designs for pendants, brooches and general motifs

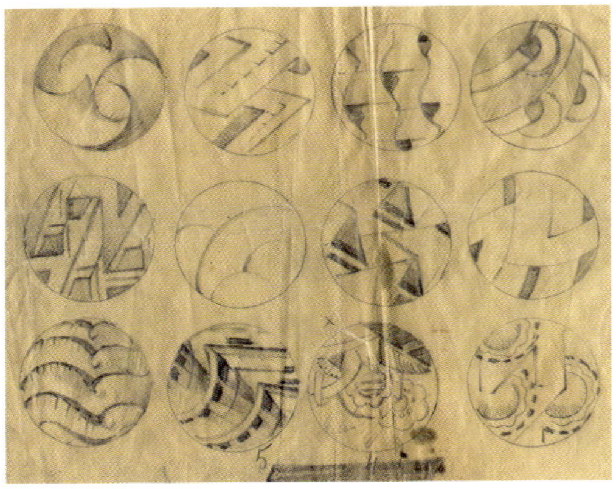

Decoration designs for vases

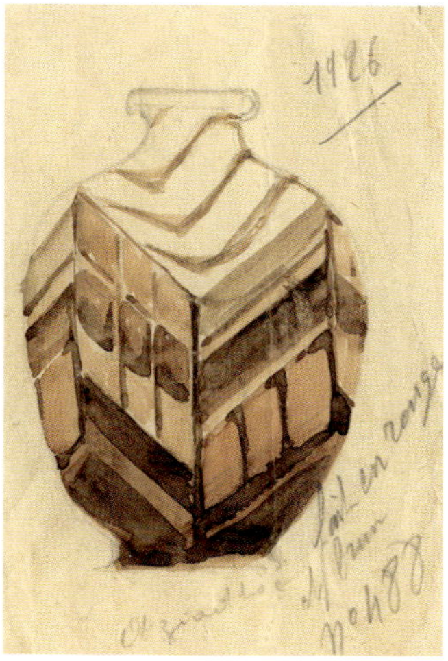

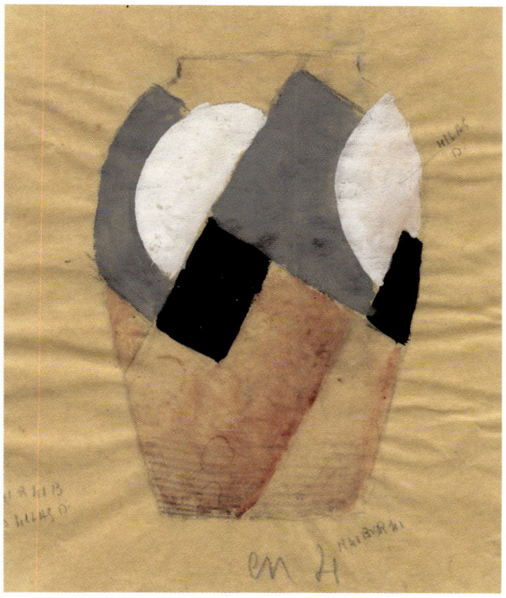

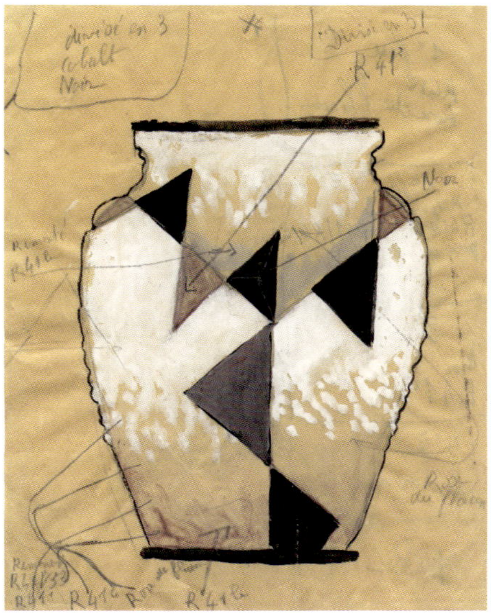

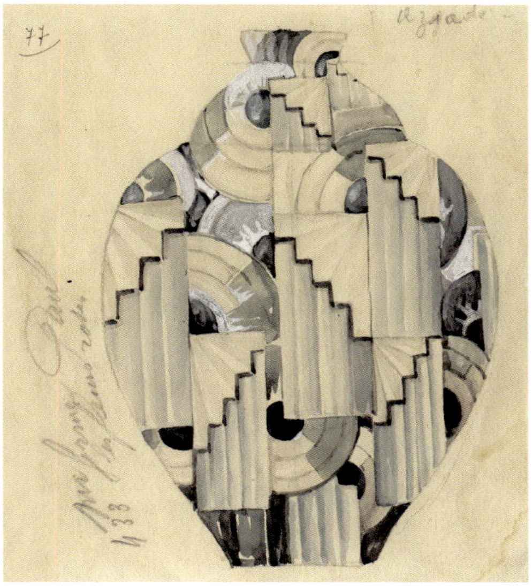

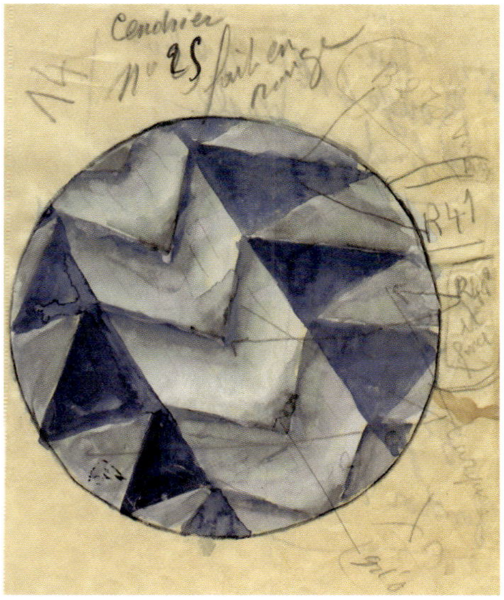

Gouache designs for vases, 1925-1935

Gouache design for vase. Hand written: "Ce qu'il y a de plus réel en moi, ce sont les illusions que je crée. Delacroix."
["What is most real about me, are the illusions that I have created."]

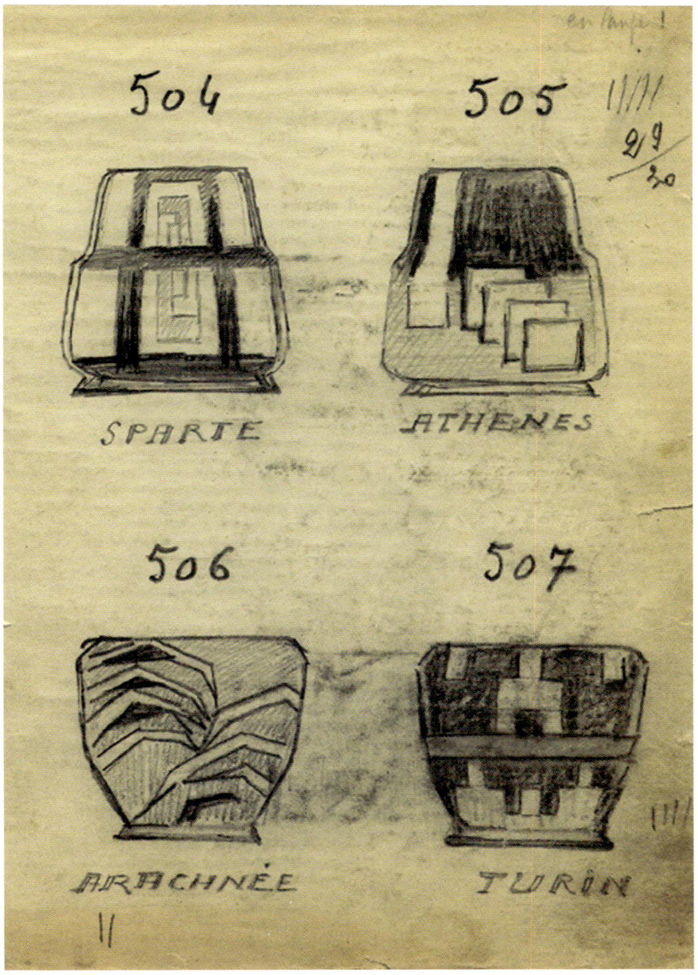

Pencil designs for vases, 1929-1930

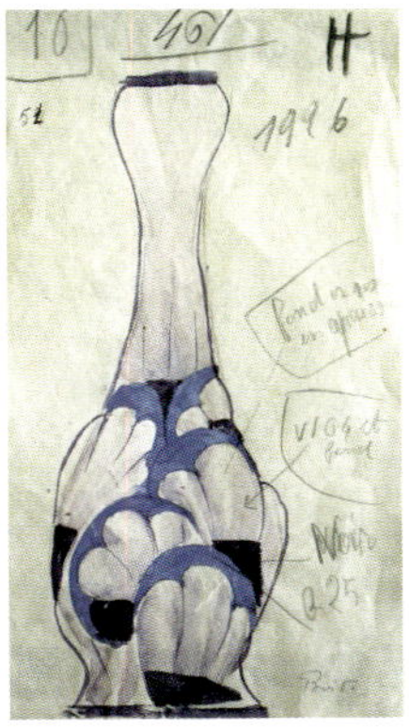

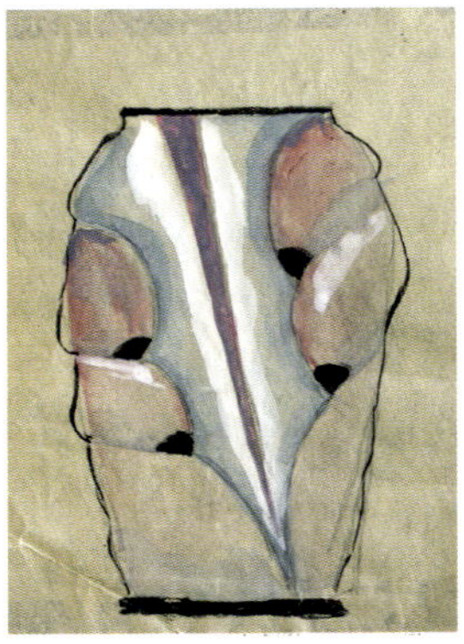

Designs for vases, 1926-1930

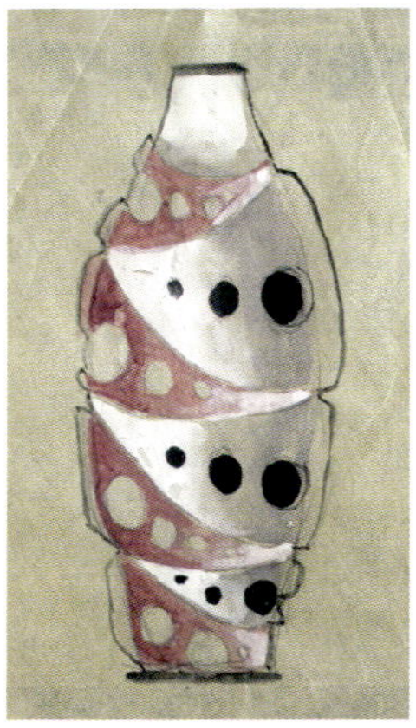

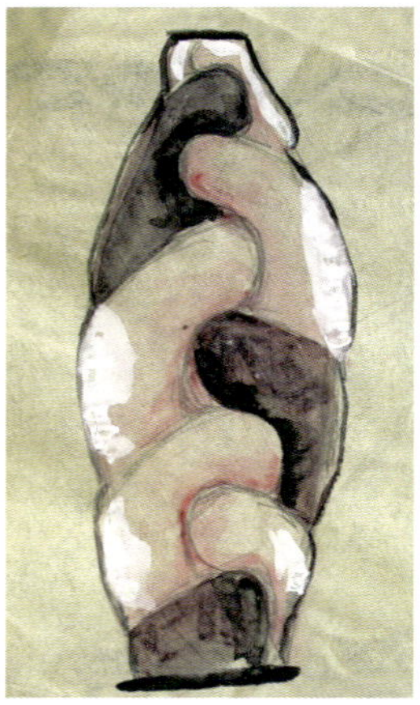

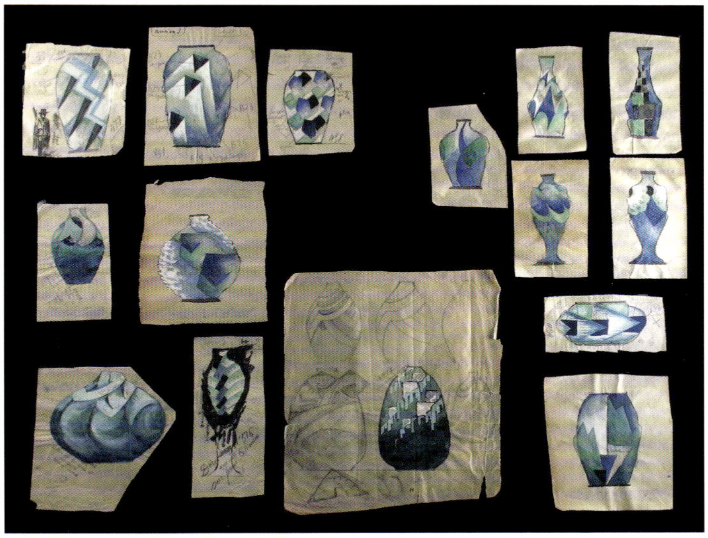

Gouache designs for clocks.
These may be prototypes as no clocks are known to have these designs

Signatures and Seals

Examples of the most typical type of signatures

Bottom of a vase

Bottom of a vase

Outer part of the neck of a vase

Inner part of a vase

Body of a vase

Back of a brooch

Plaque destined as a brooch

During Camille Fauré's life vases, plaques, jewellery and other items were signed 'C. Fauré Limoges'. This signature was applied during the final firing by different artists in charge of such a task; there was never even a thought of putting the signature of the specific artist who had actually worked on that model. The signature – either in capital letters or a mixture of capital and cursive letters – was usually in gilt paint. However, it could also be in black or dark blue, depending on what the artist found more aesthetic. Whether or not a 'dot' was added after the 'C' also varied; in most cases the word Limoges was included.

The position of the signature varies without any specific rules. On vases, if it is on the rim, it could be on the outer or inner part of it, though it is usually on the inner part. In other cases the signature is very close to the base. On plaques, the signature is usually to be found on the bottom right-hand corner. On dishes and round centrepieces it is located on the front part of the rim. On jewellery – whether brooch, pendant, bracelet or necklace – it can be either on the back of the item, or on the rim at the front.

If a piece was specifically destined for export, then the word 'France' was added as in 'C. Fauré Limoges France'. If it was destined for the U.S.A., then 'Made in France' would be added.

With the death of Camille Fauré in 1956 the signature lost the 'C' to become simply 'Fauré Limoges'. This signature was used for several years. However, in response to comments from retailers and art dealers that pieces without the letter 'C' were less in demand, the former signature was reinstated. Although this may lead to confusion as to the date of certain pieces, this can usually be clarified by looking at the design.

All the items that came out of the Fauré atelier were signed. However, some prototype pieces, such as studies and test items, were occasionally left unsigned until approved. Also one may find that the gilt signatures have been partially or completed rubbed off, due to excessive handling.

Vases for export

Gilt cursive signature

Black lettering

A piece destined specifically for the U.S.A. with the words 'Made in France'

Block letters

Fauré Marty

A Fauré Marty signature on the inner part of the neck of a vase

Fauré Marty signature on the back of a brooch

Seals

Initials FC in sealing wax

The Fauré signature. These seals appear at the back of plaques, usually attached to a label and a piece of ribbon to authenticate a piece

Labels

Émaux d'Art
du Maître Émailleur
C. FAURÉ
LIMOGES

TH 583
Marine
marée basse

Label behind mounted plaques with the name of the work of art

An 'appellation controlée' stamp. This label is found on the back of mounted plaques. It certifies that the item is from the Limousin area

The Enamels

Vase
Height: 13.5 cm. [5⅜ in.]
Signed: Fauré Marty Limoges
Private collection

Pendant
Size of enamel plaque: 2.4 cm. long [1 in.]
Signed: Fauré Marty Limoges
Hernan Salazar collection, Las Vegas

Vase
Height: 10.5 cm. [4⅛ in.]
Signed: Fauré Marty Limoges Orfebreria Anezin
Private collection

Opposite: **Vase**
Height: 20.3 cm. [8 in.]
Signed: C Fauré Marty Limoges
Liza Feldman collection

Vase
Height: 23.9 cm. [9⅜ in.]
Signed: Fauré Marty Limoges
Musée Municipal de l'Evêché, Limoges

Vase
Height: 14.3 cm. [5⅝ in.]
Signed: Fauré Marty Limoges
Private collection

Dish and vase

Dish: Height: 3.5 cm. [1⅜ in.] Diameter: 15 cm. [5⅞ in.]
Vase: Height: 14.5cm. [5¾ in.]
Both signed: Fauré Limoges France
Liza Feldman collection

Vase

Height: 14.6 cm. [5¾in.]
Signed: Fauré Marty Limoges
Hernan Salazar collection, Las Vegas

Vase
Height: 10.8 cm. [4¼ in.]
Signed: C Fauré Limoges France
Liza Feldman collection

Vase
Height: 10.8 cm. [4¼ in.]
Signed: C Fauré Limoges
Liza Feldman collection

Vase
Height: 11 cm. [4⅜ in.]
Signed: C Fauré Limoges
Private collection

Vase
Height: 10.8 cm. [4¼ in.]
Signed: C Fauré Limoges France
Private collection

Vase
Height: 9.5 cm. [3¾ in.]
Signed: C Fauré Limoges
Liza Feldman collection

Vase
Height: 9.7 cm. [3⅞ in.]
Signed: C Fauré Limoges
Liza Feldman collection

Vase
Height: 9.7 cm. [3⅞ in.]
Signed: Fauré Limoges France
Liza Feldman collection

Vase
Height: 9.7 cm. [3⅞ in.]
Signed: C Fauré Limoges
Private collection

Vase
Height: 13.5 cm. [5⅜ in.]
Signed: C Fauré Limoges
Private collection

Vase
Height: 14 cm. [5½ in.]
Signed: C Fauré Limoges France
Private collection

Vase
Height: 14.3 cm. [5⅝ in.]
Signed: C Fauré Limoges France
Private collection

Vase
Height: 9.7 cm. [3⅞ in.]
Signed: C Fauré Limoges France
Private collection

Vase
Height: 9.7 cm. [3⅞ in.]
Signed: C Fauré Limoges France
Private collection

Vase
Height: 9.7 cm. [3⅞ in.]
Signed: Fauré Limoges France
Private collection

Vase
Height: 11.5 cm. [4½ in.]
Signed: C Fauré Limoges France
Private collection

Vase
Height: 11.5 cm. [4½ in.]
Signed: C Fauré Limoges France
Private collection

Vase
Height: 11.5 cm. [4½ in.]
Signed: C Fauré Limoges France
Private collection

Vase
Height: 11.5 cm. [4½ in.]
Signed: C Fauré Limoges
Private collection

Vase
Height: 9 cm. [3½ in.]
Signed: C Fauré Limoges France
Liza Feldman collection

Vase
Height: 9.2 cm. [3⅝ in.]
Signed: Fauré Limoges France
Private collection

Vase
Height: 9.2 cm. [3⅝ in.]
Signed: C Fauré Limoges France
Private collection

Vase
Height: 9 cm. [3½ in.]
Signed: Fauré Limoges France
Private collection

Vase
Height: 9.2 cm. [3⅝ in.]
Signed: C Fauré Limoges
Private collection

Vase
Height: 12 cm. [4¾ in.]
Signed: C Fauré Limoges
Private collection

Opposite: **Vase**
Height: 14.5 cm. [5¾ in.]
Signed: C Fauré Limoges
Private collection

Vase
Height: 7.5 cm. [3 in.]
Signed: C Fauré Limoges
Private collection

Vase
Height: 7.5 cm. [3 in.]
Signed: C Fauré Limoges France
Private collection

Vase
Height: 11.4 cm. [4½ in.]
Signed: C Fauré Limoges France
Private collection

Vase
Height: 13.5 cm. [5⅜ in.]
Signed: C Fauré Limoges France
Liza Feldman collection

Vase
Height: 13.5 cm. [5⅜ in.]
Signed: C Fauré Limoges
Private collection

Vase
Height: 13.5 cm. [5⅜ in.]
Signed: C Fauré Limoges
Private collection

Vase
Height: 13.5 cm. [5⅜ in.]
Signed: C Fauré Limoges
Private collection

Vase
Height: 13.5 cm. [5⅜ in.]
Signed: Fauré Limoges France
Private collection

Vase
Height: 11.8 cm. [4⅝ in.]
Signed: C Fauré Limoges
Private collection

Vase
Height: 9.7 cm. [3⅞ in.]
Signed: Fauré Limoges France
Private collection

Vase
Height: 10.3 cm. [4 in.]
Signed: C Fauré Limoges
Private collection

Vase
Height: 10 cm. [4 in.]
Signed: C Fauré Limoges
Private collection

Vase
Height: 9 cm. [3½ in.]
Signed: C Fauré Limoges
Private collection

Vase
Height: 14.5 cm. [5¾ in.]
Signed: C Fauré Limoges France
Private collection

Vase
Height: 23.5 cm. [9¼ in.]
Signed: C Fauré Limoges
Private collection

Vase
Height: 27.2 cm. [10¾ in.]
Signed: C Fauré Limoges
Liza Feldman collection

Opposite: **Vase**
Height: 15 cm. [5⅞ in.]
Signed: C Fauré Limoges
Private collection

Vase
Height: 15.2 cm. [6 in.]
Signed: C Fauré Limoges
Liza Feldman collection

Vase
Height: 15 cm. [5⅞ in.]
Signed: C Fauré Limoges
Liza Feldman collection

Vase
Height: 11.8 cm. [4⅝ in.]
Signed: C Fauré Limoges
Liza Feldman collection

Vase
Height: 25 cm. [9⅞ in.]
Signed: C Fauré Limoges
Liza Feldman collection

Vase
Height: 11.7 cm. [4⅝ in.]
Signed: C Fauré Limoges France
Private collection

Vase
Height: 17.5 cm. [6⅞ in.]
Signed: Fauré Limoges France
Private collection

Vase
Height: 29 cm. [11⅜ in.]
Signed: C Fauré Limoges
Liza Feldman collection

Vase
Height: 29 cm. [11⅜ in.]
Signed: C Fauré Limoges
Private collection

Vase
Height: 30 cm. [11¾ in.]
Signed: C Fauré Limoges France
Private collection

Vase
Height: 17.5 cm. [6⅞ in.]
Signed: C Fauré Limoges
Private collection

Vase
Height: 13.3 cm. [5¼ in.]
Signed: C Fauré Limoges
Private collection

Vase
Height: 20 cm. [7⅞ in.]
Signed: C Fauré Limoges France
Private collection

Vase
Height: 24.5 cm. [9⅝ in.]
Signed: C Fauré Limoges
Private collection

Vase
Height: 30 cm. [11¾ in.]
Signed: C Fauré Limoges
Private collection

Vase
Height: 14.5 cm. (5¾ in.]
Signed: C Fauré Limoges
Private collection

Vase
Height: 24.5 cm. [9⅝ in.]
Signed: C Fauré Limoges
Private collection

Vase
Height: 20 cm. [7⅞ in.]
Signed: Fauré Limoges France
Private collection

Vase
Height: 15.5 cm. [6⅛ in.]
Signed: Fauré Limoges France
Private collection

Vase
Height: 12 cm. [4¾ in.]
Signed: C Fauré Limoges France
Private collection

Vase
Height: 12.5 cm. [4⅞ in.]
Signed: C Fauré Limoges
Private collection

Vase
Height: 8.3 cm. [3¼ in.]
Signed: C Fauré Limoges France
Private collection

Vase
Height: 24 cm. [9½ in.]
Signed: C Fauré Limoges France
Liza Feldman collection

Vase
Height: 15.5 cm. [6⅛ in.]
Signed: C Fauré Limoges
Private collection

Vase
Height: 20.5 cm. [8⅛ in.]
Signed: C Fauré Limoges
Private collection

Vase
Height: 26 cm. [10¼ in.]
Signed: C Fauré Limoges
Private collection

Vase
Height: 24.7 cm. [9¾ in.]
Signed: C Fauré Limoges
Private collection

Vase
Height: 24.3 cm. [9⅝ in.]
Signed: C Fauré Limoges
Private collection

Vase
Height: 27.5 cm. [10⅞ in.]
Signed: C Fauré Limoges
Liza Feldman collection

Vase
Height: 20 cm. [7⅞ in.]
Signed: C Fauré Limoges
Private collection

Vase
Height: 20 cm. [7⅞ in.]
Signed: C Fauré Limoges
Private collection

Vase
Height: 17 cm. [6¾ in.]
Signed: C Fauré Limoges
Liza Feldman collection

Vase
Height: 17.7 cm. [7 in.]
Signed: C Fauré Limoges
Liza Feldman collection

Vase
Height: 23.5 cm. [9¼ in.]
Signed: C Fauré Limoges
Liza Feldman collection

Vase
Height: 24.5 cm. [9⅝ in.]
Signed: C Fauré Limoges
Liza Feldman collection

Vase
Height: 11.3 cm. [4½ in.]
Signed: C Fauré Limoges
Liza Feldman collection

Vase
Height: 20.3 cm. [8 in.]
Signed: C Fauré Limoges France
Private collection

Vase
Height: 12.5 cm. [4⅞ in.]
Signed: Fauré Limoges France
Private collection

Vase
Height: 17 cm. [6¾ in.]
Signed: C Fauré Limoges France
Liza Feldman collection

Vase
Height: 18.3 cm. [7¼ in.]
Signed: C Fauré Limoges France
Private collection

Vase
Height: 21.5 cm. [8½ in.]
Signed: Fauré Limoges France
Private collection

Vase
Height: 13.5 cm. [5⅜ in.]
Signed: C Fauré Limoges France
Private collection

Vase
Height: 29 cm. [11⅜ in.]
Signed: Fauré Limoges France
Label on bottom reads: Ayzadée no 69301 Fleurs Chinoises
Private collection

Vase
Height: 20 cm. [7⅞ in.]
Signed: C Fauré Limoges
Liza Feldman collection

Vase
Height: 25 cm. [9⅞ in.]
Signed: C Fauré Limoges
Musée Municipal de l'Evêché, Limoges

Vase
Height: 17 cm. [6¾ in.]
Signed: C Fauré Limoges
Private collection

Vase
Height: 28.7 cm. [11¼ in.]
Signed: C Fauré Limoges
Private collection

Vase
Height: 18 cm. [7⅛ in.]
Signed: C Fauré Limoges
Private collection

Vase
Height: 17.8 cm. [7 in.]
Signed: C Fauré Limoges France
Private collection

Vase with cover
Height: 39 cm. [15⅜ in.]
Signed: Fauré Limoges France
Private collection

Vase with cover
Height: 38.5 cm. [15⅛ in.]
Signed: C Fauré Limoges
Private collection

Vase with cover
Height: 39.5 cm. [15½ in.]
Signed: Fauré Limoges France
Private collection

Vase
Height: 30.5 cm. [12 in.]
Signed: C Fauré Limoges
Private collection

Vase
Height: 30 cm. [11¾ in.]
Signed: C Fauré Limoges
Private collection

Vase
Height: 31 cm. [12¼ in.]
Signed: Fauré Limoges France
Liza Feldman collection

Back view

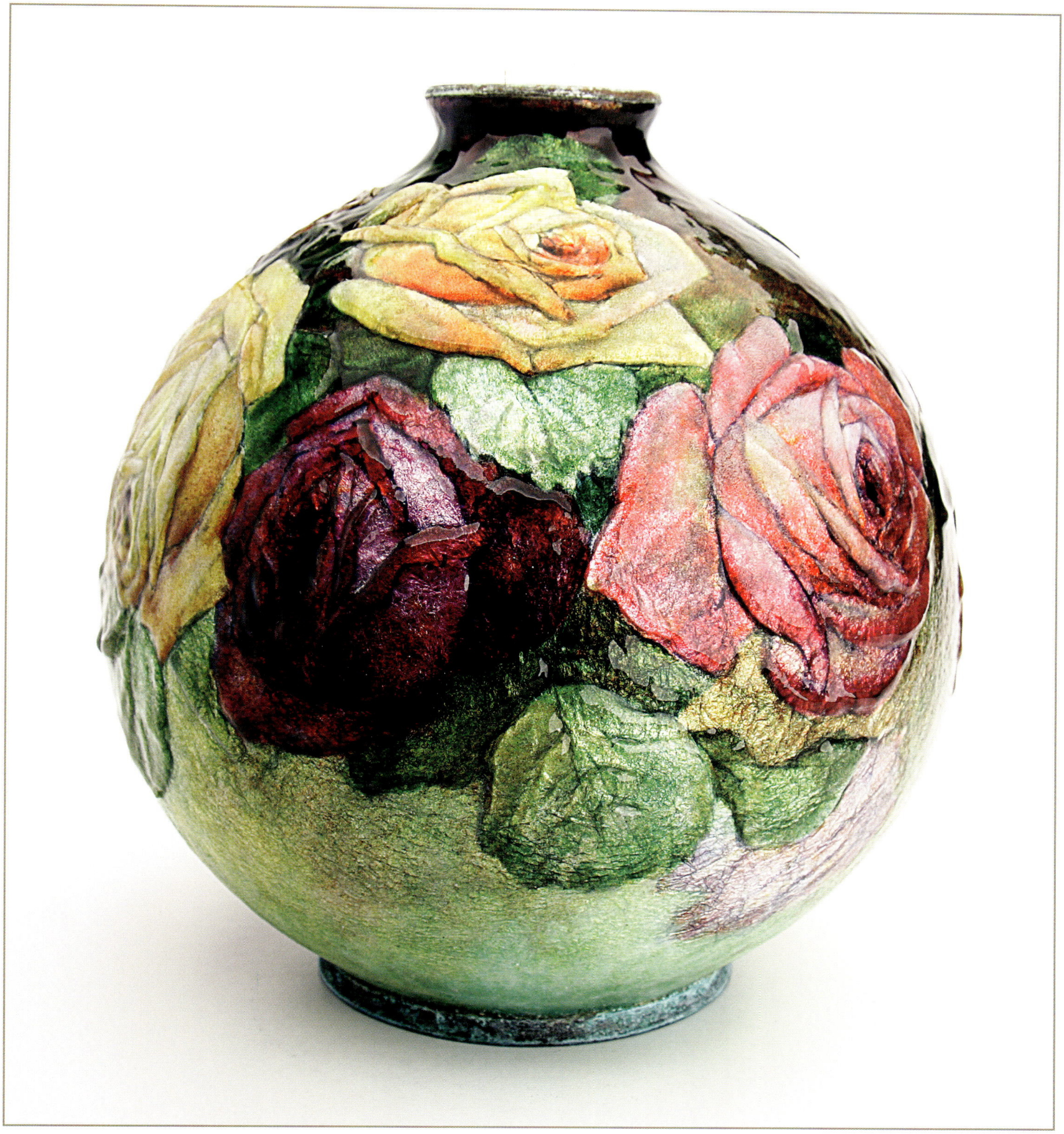

Vase
Height: 20.5 cm. [8⅛ in.]
Signed: C Fauré
Private collection

Vase
Height: 31 cm. [12¼ in.]
Signed: C Fauré Limoges
Private collection

Vase
Height: 16.8 cm. [6⅝ in.]
Signed: C Fauré Limoges
Liza Feldman collection

Vase
Height: 17 cm. [6¾ in.]
Signed: C Fauré Limoges
Musée Municipal de l'Evêché, Limoges

Vase
Height: 9.5 cm. [3¾ in.]
Signed: C Fauré Limoges
Private collection

Vase
Height: 17.5 cm. [6⅞ in.]
Signed: C Fauré Limoges
Private collection

Vase
Height: 7.3 cm. [2⅞ in.]
Signed: C Fauré Limoges
Private collection

Vase
Height: 14.5 cm. [5¾ in.]
Signed: C Fauré Limoges
Private collection

Vase
Height: 24 cm. [9½ in.]
Signed: C Fauré Limoges
Private collection

Vase
Height: 24 cm. [9½ in.]
Signed: C Fauré Limoges
Private collection

Vase
Height: 24.5 cm. [9⅝ in.]
Signed: C Fauré Limoges
Private collection

Vase
Height: 29 cm. [11⅜ in.]
Signed: C Fauré Limoges
Private collection

Vase
Height: 29.2 cm. [11½ in.]
Signed: C Fauré Limoges
Private collection

Vase
Height: 29.2 cm. [11½ in.]
Signed: C Fauré Limoges
Private collection

Vase
Height: 25.5 cm. [10 in.]
Signed: C Fauré Limoges
Private collection

Vase
Height: 25.7 cm. [10⅛ in.]
Signed: C Fauré Limoges
Private collection

Vase
Height: 31 cm. [12¼ in.]
Signed: C Fauré Limoges
Private collection

Vase
Height: 25.5 cm. [10 in.]
Signed: C Fauré Limoges
Private collection

Vase
Height: 30.5 cm. [12 in.]
Signed: C Fauré Limoges
Private collection

Vase
Height: 29.5 cm. [11⅝ in.]
Signed: C Fauré Limoges
Private collection

Vase
Height: 20 cm. [7⅞ in.]
Signed: C Fauré Limoges
Private collection

Vase
Height: 20.3 cm. [8 in.]
Signed: C Fauré Limoges
Private collection

Vase
Height: 20.3 cm. [8 in.]
Signed: C Fauré Limoges
Private collection

Vase
Height: 29 cm. [11⅜ in.]
Signed: C Fauré Limoges
Private collection

Vase
Height: 23 cm. [9 in.]
Signed: C Fauré Limoges
Private family collection

Vase
Height: 19 cm. [7½ in.]
Signed: C Fauré Limoges
Private family collection

Vase
Height: 18 cm. [7⅛ in.]
Signed: C Fauré Limoges
Private family collection

Vase
Height: 12.3 cm. [4⅞ in.]
Signed: C Fauré Limoges
Liza Feldman collection

Vase
Height: 26 cm. [10¼ in.]
Signed: C Fauré Limoges
Fondazione "Il Vittoriale degli Italiani"- Gardone Riviera (BS), Italy

Vase
Height: 16 cm. [6¼ in.]
Signed: C Fauré Limoges
Fondazione "Il Vittoriale degli Italiani"- Gardone Riviera (BS), Italy

Vase
Height: 18 cm. [7⅛ in.]
Signed: C Fauré Limoges
Fondazione "Il Vittoriale degli Italiani"- Gardone Riviera (BS), Italy

Vase
Height: 14.5 cm. [5¾ in.]
Signed: C Fauré Limoges
Private collection

Opposite: **Vase**
Height: 27.5 cm. [10⅞ in.]
Signed: C Fauré Limoges
Andy Thomain collection, Limoges

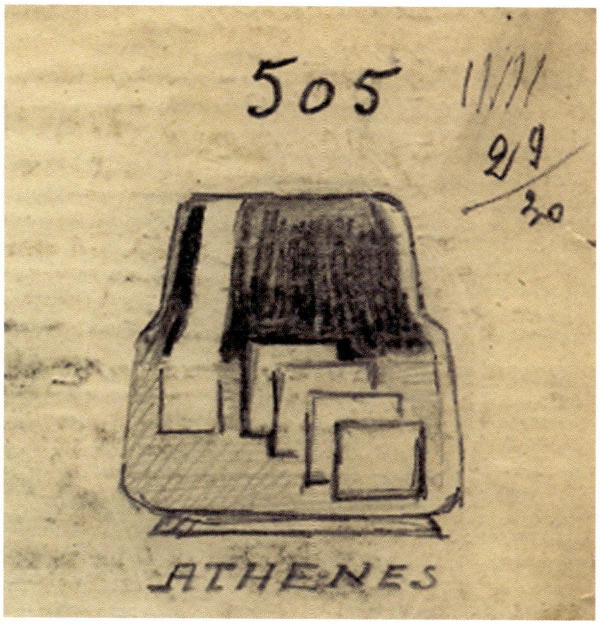

Athènes design on inverted Fausta form.
1929-30.

Vase
Height: 22 cm. [8⅝ in.]
Signed: C Fauré Limoges
Private collection

Vase
Height: 29 cm. [11⅜ in.]
Signed: C Fauré Limoges. 11 novembre 1941 Souvenir du Personnel du Commissariat Spécial de Limoges à M. Merlen.
Musée Municipal de l'Evêché, Limoges

Vase
Height: 29.5 cm. [11⅝ in.]
Signed: C Fauré Limoges
Private Collection, Rio de Janeiro

Vase
Height: 20 cm. [7⅞ in.]
Signed: C Fauré Limoges France
Private collection

Vase
Height: 18 cm. [7⅛ in.]
Signed: C Fauré Limoges
Private collection

Vase
Height: 29 cm. [11⅜ in.]
Signed: C Fauré Limoges
Made in France
Private collection

Vase
Height: 25 cm. [9⅞ in.]
Signed: C Fauré Limoges France
Private collection

Vase
Height: 30 cm. [11¾ in.]
Signed: C Fauré Limoges
Private collection

Vase
Height: 25 cm. [9⅞ in.]
Signed: C Fauré Limoges France
Private collection

Vase
Height: 30 cm. [11¾ in.]
Signed: C Fauré Limoges
Private collection

Vase
Height: 31 cm. [12¼ in.]
Signed: C Fauré Limoges France
Private collection

Vase
Height: 12.5 cm. [4⅞ in.]
Signed: C Fauré Limoges
Private collection

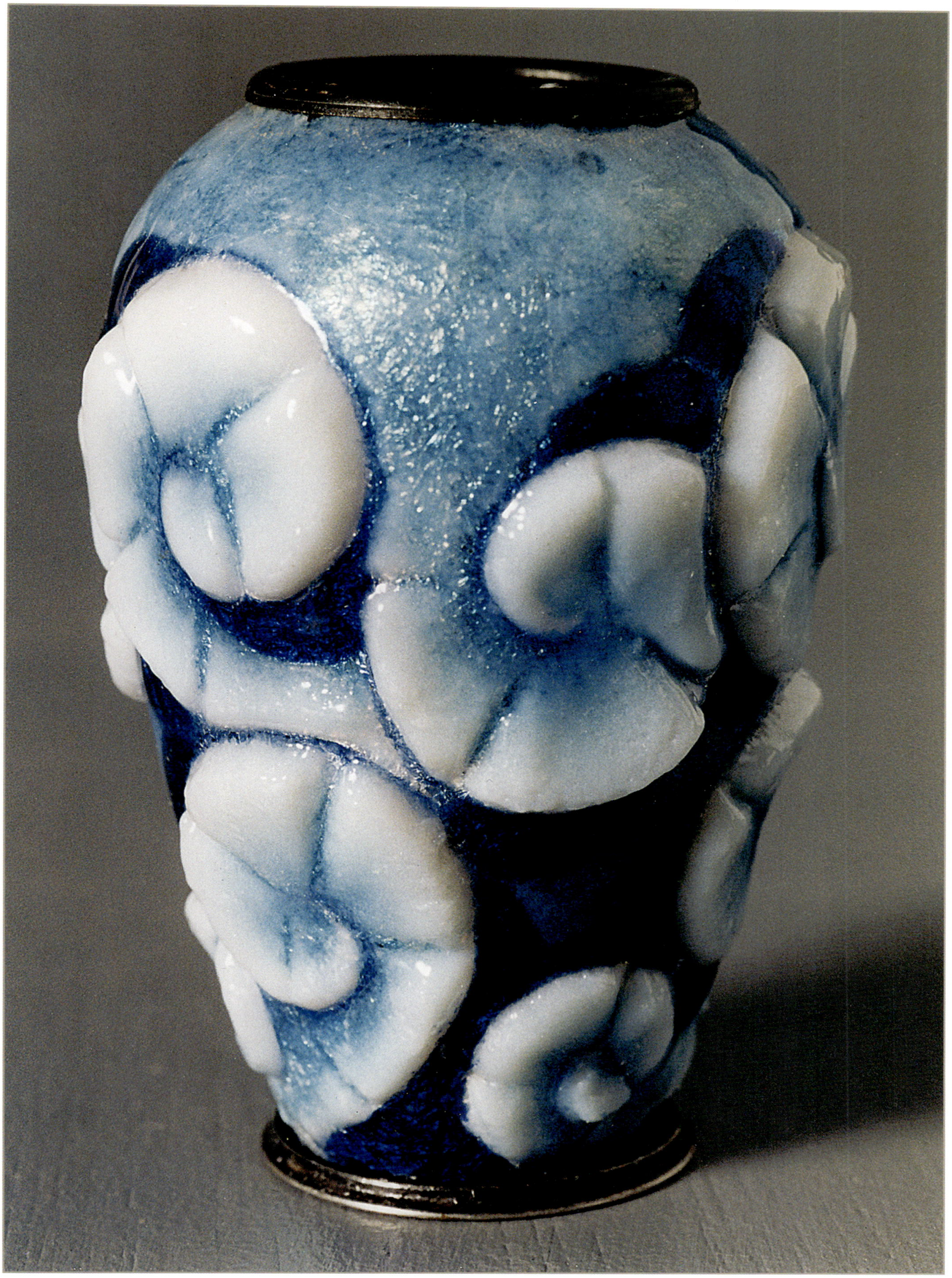

Vase
Height: 22 cm. [8⅝ in.]
Signed: C Fauré, Limoges
Andy Thomain collection, Limoges

Vase
Height: 11 cm. [4⅜ in.]
Signed: C Fauré Limoges
Private family collection, Marseille

Vase
Height: 20.3 cm. [8 in.]
Signed: C Fauré Limoges France
Private collection

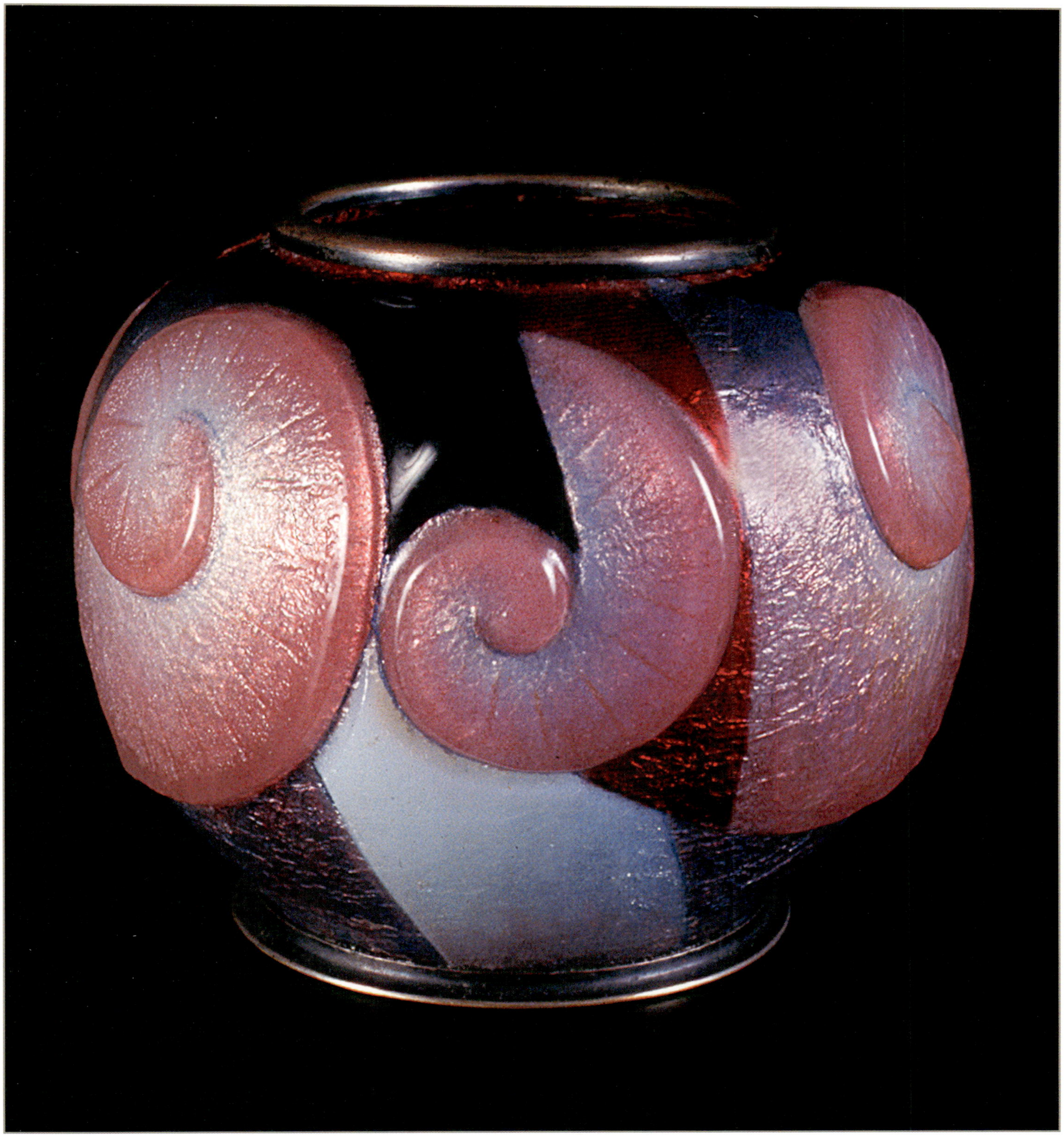

Vase
Height: 9.8 cm. [3⅞ in.]
Signed: C Fauré, Limoges
Musée Municipal de l'Evêché, Limoges

Vase
Height: 18.4 cm. [7¼ in.]
Signed: C Fauré Limoges France
Private collection

Vase used as a lamp base, contemporary photograph

Design for vase, 1927

Vase
Height: 30 cm. [11¾ in.]
Signed: C Fauré Limoges
Musée des Arts Décoratifs, Paris

C. Fauré Limoges

Vase
Height: 18 cm. [7⅛ in.]
Signed: C Fauré Limoges Made in France
Private collection

Vase with cover
Height: 29.2 cm. [11½ in.]
Signed: C Fauré Limoges
Private collection

Design for vase, 1926-1930

Vase
Height: 9 cm. [3½ in.]
Unsigned
Private collection

Vase with cover
Height: 40 cm. [15¾in.]
Signed: C Fauré Limoges
Musée de l'Horlogerie et de l'Emaillerie, Genève

Vase
Height: 24 cm. [9½ in.]
Signed: C Fauré Limoges
Private collection

Vase with cover
Height: 38 cm. [15 in.]
Signed: C Fauré Limoges France
Private collection

Vase
Height: 16 cm. [6¼ in.]
Signed: C Fauré, Limoges
Private collection

Vase
Height: 18 cm. [7⅛ in.]
Signed: C Fauré, Limoges
Private collection

Vase
Height: 16 cm. [6¼ in.]
Signed: C Fauré Limoges
Private family collection

Vase
Height: 12.5 cm. [4⅞ in.]
Signed: C Fauré Limoges
Private collection

Vase
Height: 18 cm. [7⅛ in.]
Signed: C Fauré Limoges
Private collection

Vase with cover
Height: 27.3 cm. [10¾ in.]
Signed: C Fauré Limoges France
Private collection

Vase
Size: 11cm. [4⅜ in.]
Signed: C Fauré Limoges
Private family collection

Vase
Height: 14 cm. [5½ in].
Signed: C Fauré Limoges
Private family collection

Vase
Height: 18 cm. [7⅛ in.]
Signed: C Fauré Limoges France
Private collection

Vase
Height: 9 cm. [3½ in.]
Signed: C Fauré Limoges
Musée Municipal de l'Evêché, Limoges

Vase
Height: 14.5 cm. [5¾ in.]
Signed: C Fauré Limoges
Private collection

Vase
Height: 18 cm. [7⅛ in.]
Signed: Fauré Limoges France
Private collection

Vase
Height: 33 cm. [13 in.]
Signed: C Fauré, Limoges
Private family collection, Limoges

Vase
Height: 25 cm. [9⅞ in.]
Signed: C Fauré Limoges France
Private collection

Vase with cover
Height: 23.5 cm. [9¼ in.]
Signed: C Fauré Limoges France
Private collection

Vase
Height: 24 cm. [9½ in.]
Signed: C Fauré Limoges France
Private collection

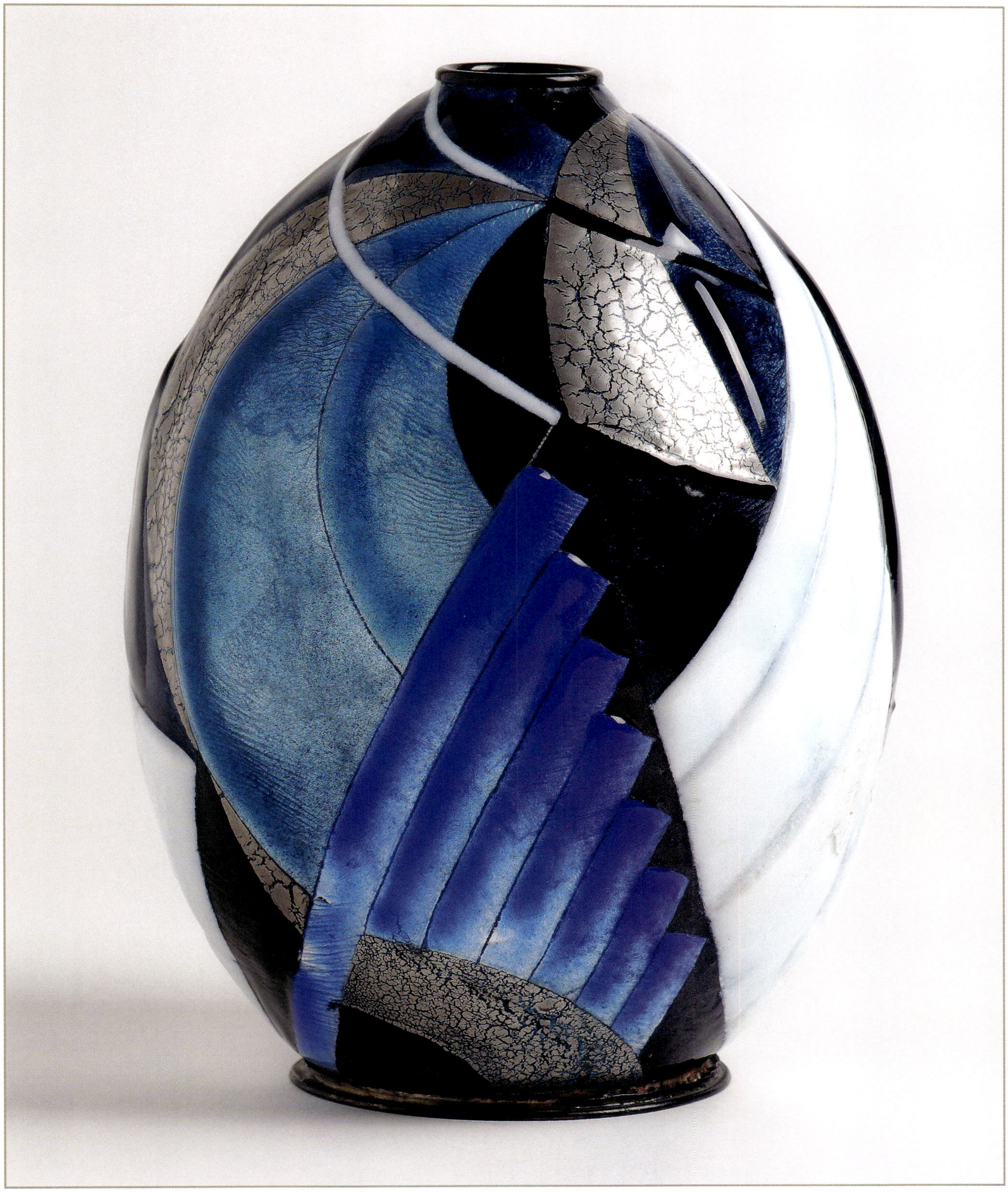

Vase
Size: 29 cm. [11⅜ in.]
Signed: C Fauré Limoges
Private family collection

Vase
Height: 29.5 cm. [11⅝ in.]
Unsigned.
This is an example of how the enamel would 'slide' if an item was fired for too long. A piece such as this would usually have been plunged into acid to destroy the enamel and to recuperate the copper form. This vase was given to a family member as a scuvenir. Never to be sold, it was not signed.
Private family collection

Vase
Height: 16 cm. [6¼ in.]
Unsigned
Private family collection

Opposite: **Vase**
Size: 28 cm. [11 in.]
Unsigned.
This is an example of how the enamel would 'slide' if an item was fired for too long. A piece such as this would usually have been plunged into acid to destroy the enamel and to recuperate the copper form. This vase was given to a family member as a souvenir. Never to be sold, it was not signed.
Private family collection

Vase
Height: 12.7 cm. [5 in.]
Signed: C Fauré Limoges France
Private collection

Vase
Height: 30 cm. [11¾ in.]
Signed: C Fauré
Private collection

Vase with cover
Height: 33.5cm. [13⅛ in.]
Signed: C Fauré Limoges France
Private collection

Back view

Vase
Height: 13.5 cm. [5⅜ in.]
Signed: C Fauré Limoges
Private collection

Vase
Height: 22 cm. [8⅝ in.]
Signed: C Fauré Limoges
Andy Thomain collection, Limoges

Vase
Height: 20.2 cm. [8 in.]
Signed: C Fauré Limoges Made in France
Private collection

Box
Height: 6 cm. [2⅜ in.] Diameter: 18 cm. [7⅛ in.]
Signed: C Fauré Limoges
Private family collection

Vase
Height: 27.5 cm. [10⅞ in.]
Signed: C Fauré Limoges France
Liza Feldman collection

Box
Height: 22.5 cm. [8⅞ in.]
Signed: C Fauré Limoges
Private family collection, Limoges

Design for a bonbonnière, 1952

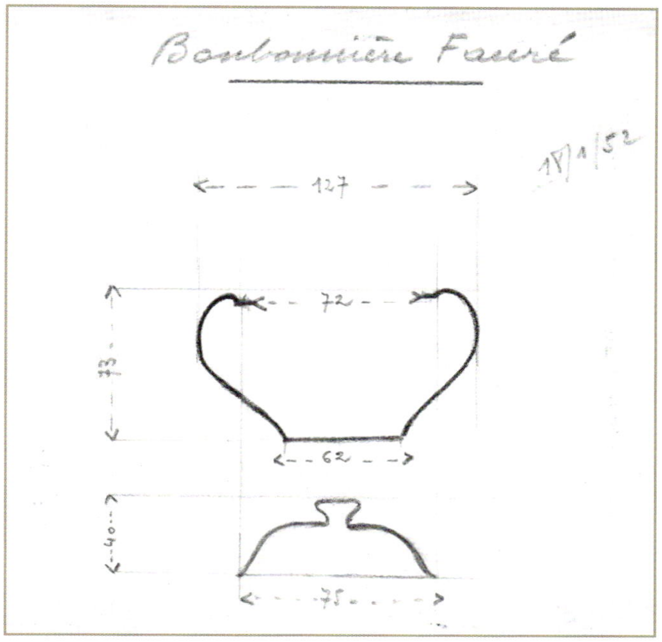

Bonbonnière
Height: 11.5 cm. [4½ in.]
Signed: C Fauré Limoges
Liza Feldman collection

Box
Height: 7 cm. [2¾ in.]
Signed: C Fauré Limoges
Andy Thomain collection, Limoges

Box
Height: 10.5 cm. [4⅛ in.] Diameter: 21 cm. [8¼ in.]
Signed: C Fauré Limoges
Private collection

Box
Height: 8.5 cm. [3⅜ in.]
Signed: C Fauré Limoges France
Liza Feldman collection

Box
Height: 7.5 cm. [3 in.] Diameter: 16.5 cm. [6½ in.]
Signed: C Fauré Limoges France
Private collection

Box
Height: 6.5 cm. [2½ in.]
Signed: C Fauré Limoges France
Andy Thomain collection, Limoges

Lamp base
Height: 24.5 cm. [9⅝ in.]
Signed: Fauré Limoges France
Private collection

FAURÉ . LIMOGES . FRANCE

Lamp base
Height: 21 cm. [8¼ in.]
Signed: C Fauré, Limoges
Private family collection, Limoges

Lamp base
Height: 32 cm. [12⅝ in.]
Signed: C Fauré, Limoges France
Private collection

Lamp base
Height: 14 cm. [5½ in.]
Signed: C Fauré Limoges
Private family collection, Limoges

Lamp
Height of lamp base: 17 cm. [6¾ in.]
Signed: C Fauré Limoges
Private collection

Lamp base with period shade by Georges Cheyron
Height of base: 13.5 cm. [5⅜ in.]
Unsigned
Private family collection, Limoges

Lamp base with period shade by Georges Cheyron
Height of base: 19 cm. [7½ in.]
Signed: C Fauré Limoges
Private family collection, Limoges

Dish
Height: 3.5cm. [1⅜ in.] Diameter: 15 cm. [5⅞ in.]
Signed: C Fauré Limoges
Private collection

Dish
Height: 6.5 cm. [2½ in.] Diameter: 27.5 cm. [10⅞ in.]
Signed: Fauré Limoges France
Liza Feldman collection

Centrepiece
Height: 5.5 cm. [2⅛ in.] Diameter: 18 cm. [7⅛ in.]
Signed: Fauré Limoges
Private collection

Dish
Height: 3.5 cm. [1⅜ in.] Diameter: 20.5cm. [8⅛ in.]
Signed: C Fauré Limoges
Private collection

Dish
Height: 5 cm. [2 in.] Diameter: 18 cm. [7⅛ in.]
Signed: C Fauré Limoges France
Liza Feldman collection

Dish
Height: 5 cm. [2 in.] Diameter: 18 cm. [7⅛ in.]
Signed: C Fauré Limoges France
Private collection

Dish
Height: 3 cm. [1⅛ in.] Diameter: 15 cm. [5⅞ in.]
Signed: C Fauré Limoges
Private collection

Dish
Height: 3 cm. [1⅛ in.] Diameter: 15cm. [5⅞ in.]
Signed: C Fauré Limoges
Private collection

Plaque
Diameter: 22.5 cm. [8⅞ in.]
Signed: C Fauré Limoges France
Private family collection, Marseille

Plaque
Diameter: 29 cm. [11⅜ in.]
Signed: C Fauré Limoges
Private family collection

Centrepiece
Height: 5.5 cm. [2⅛ in.] Diameter: 27 cm. [10⅝ in.]
Signed: C Fauré Limoges
Musée Municipal de l'Evêché, Limoges

Centrepiece
Height: 9.2cm. [3⅝ in.] Diameter: 25.5 cm. [10 in.] without metal mount.
Signed: C Fauré Limoges
Musée Municipal de l'Evêché, Limoges

Dish
Length: 14.2 cm. [5⅝ in.]
Signed: C Fauré Limoges
Private family collection, Limoges

Dish
Height: 4.5 cm. [1¾ in.] Diameter: 22 cm. [8⅝ in.]
Signed: C Fauré Limoges
Private collection

Coaster
Diameter: 8.7 cm. [3⅜ in.]
Signed: C Fauré Limoges
Private collection

Dish
Diameter: 10.5 cm. [4⅛ in.]
Signed: C Fauré Limoges
Private family collection, Limoges

Cabinet (maker unknown)
Size of each plaque: 17.5 x 17.5 cm. [6⅞ x 6⅞ in.]
Signed: C Fauré
Private family collection

Perfume burner in iron mount
Height: 22.5 cm. [8⅞ in.]
Unsigned
Private family collection, Limoges

Plaque
Size: 20 x 15 cm. [7⅞ x 5⅞ in.]
Signed: C Fauré Limoges
Private family collection

Plaque
Size: 23.5 x 18.5 cm. [9¼ x 7¼ in.]
Signed: C Fauré Limoges
Private family collection

Plaque. Miss Graham (detail) after Gainsborough
Length: 29 cm. [11⅜ in.]
Signed: Fauré
Private family collection

Plaque
Length: 29 cm. [11⅜ in.]
Signed: C Fauré Limoges
Private family collection

Plaque
Size: 16 x 11.5 cm. [6¼ x 4½ in.]
Signed: Fauré Limoges
Private family collection, Limoges

Plaque
Size: 19.5 x 14.5 cm. [7⅝ x 5¾ in.]
Signed: C Fauré Limoges
Private family collection

Plaque
Size: 35 x 29 cm. [13¾ x 11⅜ in.]
Signed: C Fauré Limoges
Private family collection, Marseille

Plaque

Size: 20.5 x 31.5 cm. [8⅛ x 12⅜ in.]

Signed: Fauré Limoges France No 3/8

Private family collection, Limoges

Plaque
Size: 37 x 24 cm. [14⅝ x 9½ in.]
Signed: C Fauré Limoges
Private family collection, Marseille

Plaque. 'Virgin' after the Bouguereau painting in the Louvre Museum
Size: 21.5 x 30.5 cm. [8½ x 12 in.]
Signed: C Fauré Limoges
Private family collection, Limoges

Plaque. Lys
Size: 7 x 9 cm. [2¾ x 3½ in.]
Signed: Fauré Limoges France
Private collection

Plaque
Size: 5.5 x 6.5 cm. [2⅛ x 2½ in.]
Signed: C. Fauré Limoges
Private collection

Plaque. Rose rosée after Redouté
Size: 7 x 9 cm. [2¾ x 3½ in.]
Signed: Fauré Limoges France
Private collection

Plaque
Size: 8 x 11.5 cm. [3⅛ x 4½ in.]
Signed: Fauré Limoges France
Private collection

Plaque
Size: 11.5 x 16 cm. [4½ x 6¼ in.]
Signed: Fauré Limoges France
Private collection

Oval plaque
Length: 17cm. [6¾ in.]
Signed: Fauré Limoges France
Private family collection, Limoges

Oval plaque
Length: 17cm. [6¾ in.]
Signed: Fauré Limoges France
Private family collection, Limoges

Plaque
Diameter: 23 cm. [9 in.]
Signed: C Fauré Limoges
Private collection

Plaque
Size: 17.5 x 12 cm. [6⅞ x 4¾ in.]
Signed: C Fauré Limoges
Private family collection

Plaque. Based on a drawing by Jacquement
Size: 11.8 x 14 cm. [4⅝ x 5½ in.]
Signed: C Fauré Limoges
Private family collection, Limoges

Plaque. 'Flowers'
Size: 13.5 x 21 cm. [5⅜ x 8¼ in.]
Signed: C Fauré Limoges
Private familiy collection, Limoges

Plaque. 'La Dormeuse' after Renoir
Size: 19 x 24.2 cm. [7½ x 9½ in.]
Signed: Fauré Limoges France
Private collection

Plaque. 'Danseuses'
Length: 48 cm. [18⅞ in.]
Signed: C Fauré Limoges France
Private collection

Plaque. After Poliakoff
Size: 27 x 33 cm. [10⅝ x 13 in.]
Signed: Fauré Limoges d'après Poliakoff
Private family collection

Bust
Size: 49 x 34 cm. [19¼ x 13⅜ in.]
Signed: C Fauré Limoges
Private family collection

Plaque. Marine. Marée basse
Size: 8.5 x 13.5 cm. [3⅜ x 5⅜ in.]
Signed: Fauré Limoges France
Private collection

Plaque
Size: 7 x 9.5 cm. [2¾ x 3¾ in.]
Signed: Fauré Limoges France
Private collection

Plaque
Size: 12 x 30 cm. [4¾ x 11¾ in.]
Signed: Fauré Limoges France
Private collection

Brooch
Diameter: 3.5 cm. [1⅜ in.]
Signed: Fauré Limoges
Private collection

Brooch
Diameter: 6.5 cm. [2½ in.]
Signed: C Fauré Limoges
Private collection, São Paulo

Brooch
Size: 2.6 x 9 cm. [1 x 3½ in.]
Signed: C Fauré Limoges
Musée Municipal de l'Evêché, Limoges

Plaques for brooches
Diameter: 7 cm. each [2¾ in.]
Each signed: C Fauré Limoges
Private collection

Medallions
Length: 2.5 to 4.8 cm. [1 to 1⅞ in.]
Signed: C Fauré Limoges
Private family collection

Box cover
Length: 10.5 cm. [4⅛ in.]
Signed: C Fauré Limoges
Private collection

Plaques for brooches
Length: 8 cm. each [3⅛ in.]
Each signed: C Fauré Limoges
Private collection

Vase
Height: 31 cm. [12¼ in.]
Signed: Fauré Limoges France
Liza Feldman collection

Vase
Height: 29 cm. [11⅜ in.]
Signed: Fauré Limoges France
Liza Feldman collection

Vase
Height: 11.5 cm. [4½ in.]
Signed: Fauré Limoges France
Liza Feldman collection

Vase with cover
Height: 34 cm. [13⅜ in.]
Signed: Fauré Limoges France
Liza Feldman collection

Vase
Height: 17.5 cm. [6⅞ in.]
Signed: Fauré Limoges France
Private collection

Vase with cover
Height: 40 cm. [15¾ in.]
Signed: Fauré Limoges France
Liza Feldman collection

Bibliography

2e Salon Internationale des Industries et Art du Feu. 9-7 September 1962. Paris.

Allgemeines Kunstler-Lexicon. K.G. Saur. Band 37. Munich. Leipzig. 2003.

Archives of the Lyon Fair, SEPEL, Lyon.

Bebien, E., *Appareils Bebien* (Sales booklet). Paris. 1890.

Bernoux, G. (ed.), *Emaux sur cuivre.* Eyrolles. Paris. 2002.

Brunhammer, Y. and Tise, S., *The decorative arts in France. La Societé des Artistes Decorateurs.* Rizzoli, New York. 1990.

Dalpayrat, L., *Limoges enamels. Processes of the early Limoges enamellers.* Lechertier, Barbe & Co. Artists Colourmen, London. 1881.

'Emaux de Limoges. L'art du feu au XXe siècle', in *Exhibition catalogue* Hermitage Museum exhibition. St. Petersbourg. 16 December 2003 - 26 January 2004.

Ferrer, J. and Grandcoing, P. (ed.), *Une histoire de Limoges.* Culture & Patrimoine en Limousin. 2003.

Ferrer, J. and Notin, V., *L'art de l'émail à Limoges.* Culture et Patrimoine en Limousin. 2005.

Hiller, B., *The World of Art Deco.* The Minneapolis Institute of Arts exhibition. E.P. Dutton and Co., Inc. New York. 1971.

L'Illustration. 11 January 1930.

Kiener, M.C., *Emaux Art Deco, Biennale Internationale de l'Email Contemporain.* Limoges. 1992.

Lyon – Guide historique et artistique. Comité de patronage de l'exposition de 1914 (ed.).

Meyer, A., *L'Art de l'Email de Limoges.* 1895.

Morrison McClinton, K., *Art deco a guide for collectors.* Clarkson N. Potter, inc., New York. 1972.

Robert, K., *La Peinture sur email – Emaux peints – Emaux de Limoges.* 12 dessins de Jany-Robert et E. Leroy. 1928.

Sammlung Bröhan. Kunsthandwerk-Metal Porzellan. Berlin. 1977.

Zimmer, T., Sham's, L., Lavoute, F., and Maurand, L., (ed.), *Email Limousin.* Culture et Patrimoine en Limousin. 1996.

Photographic Credits

Marcelo Ferrari: cover; back cover; spine, p45 (centre left), p49 (top left), p60 (top: left, middle and right; centre: left and right), p61, p64 (top left; bottom right), p66, p70 (bottom: left and right), p71 (bottom right), p72, p73, p74, p75, p76 (top right; bottom: left and right), p77, p78, p79, p80 (bottom left; right), p81, p82, p83, p84, p87, p90, p91, p93, p94, p95, p96, p97, p98, p99, p100, p101, p102, p103, p104, p105, p106, p108, p109, p110, p111, p112, p114, p115, p121, p122, p124, p125, p126, p127, p130, p131, p132, p133, p134, p135, p136, p140, p141, p144, p145, p146, p147, p148, p149, p150, p151, p152, p153, p155, p156, p157, p158, p159, p160, p161, p162, p163, p169, p171 (bottom), p173, p174, p175, p176, p177, p178, p179, p180, p183, p184, p188, p189, p191 (right), p192, p193, p197, p198, p199, p202, p204, p205, p208, p209, p214, p215, p216, p217, p218 (left), p219, p224 (bottom), p225 (bottom), p227, p230, p233 (top left; bottom: left and right), p234 (top right; bottom: left and right), p254, p255, p257, p260, p261, p264, p265, p266 (top), p275.

Fauré Family Archives: p10, p12, p15, p16 (right), p19 (top), p22, p28, p29, p30 (top left), p33 (top left; bottom right), p34, p36, p39, p40, p58 (middle right; bottom right), p59 (left: top middle and bottom), p186 (right), p191 (left).

Bernard Vergne, Archives: p26, p43 (bottom), p46 (left), p47, p49 (all except top left), p50, p51, p52 (centre left and right; bottom left, middle and right), p223 (top).

Mauricette Pinoteau, Archives: p24, p27, p37, p44 (bottom), p54, p194, p195, p229.

Alain Grafeuil, Archives: p8, p21, p55, p56, p57 (top left), p59 (top right; bottom right), p171 (top), p172 (right), p186 (left).

Anne Garde and Aurelie Foussard: p46 (top right; bottom right), p48, p52, p53, p60 (bottom left; centre middle), p137, p241 (top), p266 (bottom), p267 (bottom), p269.

Private Archives, Marseille: p182, p235, p250, p252.

Frédéric Magnoux: p57 (all, except top left), p58 (all, except middle right and bottom right), p67, p128, p143, p172 (left), p185, p203, p237, p238, p267 (top).
Musée Municipal de l'Evêché, Limoges, Archives: p57 (all except top left), p58 (all except middle and bottom right).

Ali Elai: p65, p69, p70 (top left; top right), p71(top left; top right; bottom left), p76 (top left), p80 (top left), p85, p86, p88, p89, p92, p107, p113, p116, p117, p118, p119, p120, p123, p129, p138, p139, p142, p167, p220, p223 (bottom), p225 (top), p233 (top right), p234 (top left), p270, p271, p272, p273, p274.

Pascal Kirchner: p23, p45 (top right), p164, p165, p166, p196, p, p221, p222, p231, p232, p236, p239, p241 (bottom), p242, p244, p246, p248, p251, p253, p256, p258 (right), p259.

Pascal Helaine: p200, p201, p210, p211, p212, p213, p243, p245, p247, p249, p258 (left), p262, p263, p268.

SEPEL Archives, Lyon: p17 (top), p18, p19 (bottom).

Romulo Fialdini: p207.

Sylvain Aubry: p190.

Jean Tholance: p187.

Roger Vulliez: p154, p170, p181, p218 (right), p224 (top), p226.

Carolina Corral: p68, p64 (top right).

Sergio Ferraz: p240.

Marco Rapuzzi: p168 (left).

Augusto Rizza: p168 (top right; bottom right).

Francisco Roberto Oliveira Pinto: back flap.

Alberto Shayo, Archives: p11, p13, p16 (left), p17 (bottom), p20, p30 (top right; bottom), p32, p33 (top right), p35, p41, p42, p43 (top left; top right), p44 (top), p45 (top left; centre right; bottom), p52 (top left), p206, p228.

Index

Page numbers in bold denote images and/or captions.